Naturally Fractured
Reservoir Characterization

Other Books in the Series
Petroleum Geostatistics
Seismic Inversion

Coming Soon
Hydraulic Fracturing

Naturally Fractured Reservoir Characterization

Wayne Narr
Chevron Energy Technology Co.

David S. Schechter
Texas A&M U.

Laird B. Thompson
Utah State U.

Society of Petroleum Engineers

ISBN 978-1-55563-112-3

09 10 11 12 13 14 15 / 10 9 8 7 6 5 4 3 2

Society of Petroleum Engineers
222 Palisades Creek Drive
Richardson, TX 75080-2040 USA

http://store.spe.org
books@spe.org
1.972.952.9393

Preface

Considered physically or metaphorically, fracture systems form a labyrinth.

Reservoir characterization involves describing the nature of both the static (geologic) and dynamic (fluid-flow) nature of a rock. Naturally fractured reservoirs (NFRs) are particularly challenging to characterize because of the extreme heterogeneity of both their static and dynamic natures. Furthermore, few geologists or engineers have a high level of skill or familiarity with natural fracture systems. They constitute a niche specialization. However, the trend toward exploitation of lower-quality reservoir rocks and toward ever-increasing enhanced-oil-recovery efforts brings increasing numbers of petroleum-industry professionals face to face with naturally fractured reservoirs.

Fractures can make primary production from a low-quality reservoir economically feasible. On the other hand, fractures can compromise an enhanced-recovery project through channeling of injected fluids. Whether dealing with primary production or enhanced recovery, reservoir management decisions can be improved through better understanding of the natural fractures in a reservoir.

"Classic" NFR behavior is often a myth. Fracture systems do not behave according to "conventional" geostatistical methods (or at least we think they don't). Seismic methods for fracture system analysis are still in their infancy—or, at best, in their adolescence. Numerical simulation is challenged by extreme contrast of juxtaposed permeability systems.

There is no standard, generally agreed-upon workflow for characterizing naturally fractured reservoirs. This is due in part to the challenges posed by the flow heterogeneity induced by natural fractures and in part to the variability in character that fractures present. However, there are methods that may be employed to better understand and describe NFRs, and in this book we have compiled basic information about fracture systems, NFR behavior, and practical methods for exploiting NFRs.

Every NFR seems to be a learning experience from the standpoint of science, engineering, and economics. The surprises teach the lesson of humility when eventualities depart from expectations. From a more tangible perspective, some surprises come with dire economic consequences. This is a prime reason for seeking a broad understanding of the fracture labyrinth.

Acknowledgments

We would like to thank Chuck Kluth, Joao Keller, and Jairam Kamath for critically reviewing drafts of chapters of this book. They provided us with many helpful suggestions, improved its readability, and protected us (and our readers) from a few errors.

We appreciate the stimulus that Hossein Kazemi has provided to help move this project forward to publication, as well as his technical suggestions and input.

We thank Tom McMillen and Wen Chen of Chevron Corp. for approving and endorsing Wayne's participation in this publication.

Finally, we thank Jennifer Wegman for the effort and vigilance she has invested to facilitate the whole process of making our manuscript into a finished book.

Contents

Preface v

Acknowledgments vii

1. **Recognizing Naturally Fractured Reservoirs** 1
   1.1   Introduction 1
   1.2   Productivity Heterogeneity 2
   1.3   Example From Spraberry 9
   1.4   Understanding NFRs 13

2. **Geology of Natural Fracture Systems** 15
   2.1   Introduction 15
   2.2   Fractures 15
   2.3   Joints 17
   2.4   Faults 27
   2.5   Other Fracture-Related Features 29
   2.6   Relationship Among Fractures and Other Geologic Features 30
   2.7   Contemporary Stress Field 33
   2.8   Conclusions 34

3. **NFR Characterization** 35
   3.1   Introduction 35
   3.2   Detecting Fractures in the Well 35
   3.3   Indirect Fracture Indicators and Validation 42
   3.4   Orientation and Spatial Organization of Reservoir Fractures 46
   3.5   Fracture Density Distribution 50
   3.6   Using Analogs in NFR Characterization 52
   3.7   Using Dynamic Data in NFR Characterization 53
   3.8   Synthesizing All the Data 58

4. **Fluid Flow in NFRs** 61
   4.1   Introduction 61
   4.2   Fracture/Matrix Interaction 61
   4.3   Yates Field 64
   4.4   Simulation of NFRs 67

5. Case Histories of NFRs **71**
    5.1   Midale Field 71
    5.2   Spraberry Trend 75
    5.3   Ekofisk Field 84

Nomenclature **99**

References **101**

Index **107**

Chapter 1

Recognizing Naturally Fractured Reservoirs

1.1 Introduction

Fractures are the most abundant visible structural features in the Earth's upper crust. They are evident at most rock outcrops, and it is likely that most reservoirs contain some natural fractures. Yet determining whether fractures are present in sufficient quantity and extent to have a significant impact on reservoir behavior can be challenging. Whether to designate a reservoir as a naturally fractured reservoir (NFR) is a judgment based on the degree to which the fractures affect reservoir performance.

Fractures can have a profound impact on reservoir management and field economics; therefore, the earlier the influence of fractures is determined, the better. The presence of fractures should affect many aspects of reservoir management, including drilling, well completions, data collection, well placement, and enhanced-recovery schemes. They can be a critical factor when judging recoverable reserves. Consequently, a prudent approach to reservoir evaluation is an oft-cited admonition: *All reservoirs should be considered fractured until proven otherwise.* This chapter provides some suggestions for early recognition of NFRs.

Naturally fractured reservoirs are elusive systems to characterize and difficult to engineer and predict. However, because many reservoirs are naturally fractured, it is important to establish some basic criteria for recognizing when fractures are an important element in reservoir performance, and to understand the nature and performance characteristics of an NFR. Our intent in this primer is to provide basic information on how to recognize, characterize, and engineer naturally fractured reservoirs. This primer is not intended to provide a comprehensive review of the techniques and methods used in the industry, but rather to provide a starting point for geoscientists and engineers, to enable them to recognize and understand many key issues involved in working with NFRs, and to suggest some methods for evaluating and optimizing production when natural fractures are present.

An NFR is a reservoir in which fractures enhance the permeability field, thereby significantly affecting well productivity and recovery efficiency. Fractures are mechanical discontinuities or partings caused by brittle failure. They are observed across a vast range of scale, from nearly ubiquitous microcracks to multikilometer-long features. Fractures can be open, permeable pathways, or they can be permeability baffles resulting from the presence of sec-

TABLE 1.1—CLASSIFICATION OF NATURALLY FRACTURED RESERVOIRS, MODIFIED FROM NELSON (2001)		
Types of Fracture Reservoirs		
NFR Type	Definition	Examples
Type 1	Fractures provide essential porosity and permeability.	Amal, Libya Edison, California Basement fields, Kansas
Type 2	Fractures provide essential permeability	Agha Jari, Iran Haft Kel, Iran Sooner trend, Oklahoma Spraberry trend area, Texas
Type 3	Fractures provide a permeability assistance	Kirkuk, Iraq Dukhan, Qatar Cottonwood Creek, Wyoming Lacq, France

ondary mineralization or other fine-grained material filling their apertures. Fractures can assist the production of hydrocarbons or prevent its economic extraction.

Naturally fractured reservoirs have been classified according to the relative contribution of the matrix and fractures to the total fluid production. We follow the classification scheme proposed by Nelson (2001), but with modification (**Table 1.1**). Following our definition of NFRs, we treat only those reservoirs in which fractures act to enhance permeability. Fractures, especially faults, can certainly diminish permeability, but this effect is so distinct in both geologic occurrence and engineering practice that we have chosen not to consider fault-seal issues within this volume. Table 1.1 shows a modified classification of naturally fractured reservoirs based on the system proposed by Nelson (2001).

An initial analysis task, which can be a significant challenge, is posed by the simple question, "Is my reservoir an NFR?" In some cases, this question has remained unresolved for many years after discovery. The operator needs to assess early in the producing life of a reservoir whether production is affected by the presence of naturally occurring fractures. Some NFRs are identified during the initial drilling phase when severe mud losses or dramatic bit drops into solution-widened fracture cavities occur. After a reservoir has been developed, the impact of fractures may become evident as a consequence of aberrant performance—such as production interference or well productivity greatly in excess of matrix-based expectations.

1.2 Productivity Heterogeneity

In practice, the evidence that a reservoir is naturally fractured can be ambiguous. If one well in a 20-well field shows clear evidence of fracture-enhanced production from the main reservoir but none of the other 19 wells show any fracture effects, most workers would not call this an NFR. But what if 8 wells show clear evidence of fracture-enhanced flow but 12 do not? The definition of an NFR allows wide latitude for discretion. Unanimous agreement on whether to call any particular reservoir an NFR is highly unlikely. Part of the reason for this judgment problem is their significant heterogeneity, and the other part is the gradational character of the transition between "conventional" reservoirs and NFRs.

Well-production heterogeneity is a characteristic of most NFRs. Although various geologic processes other than fractures can lead to fieldwide heterogeneity, such as thin, discontinuous high-permeability strata or variable development of interconnected vugs, we can use heterogeneity as evidence of possible fracture involvement. Heterogeneity raises suspicion, but requires corroborating evidence to convict.

Beliveau (1995) examines productivity improvements in horizontal wells compared to neighboring vertical wells in a large number of reservoirs [Productivity Improvement Factor (PIF) = stable oil or gas rate compared to the current rate of a neighboring vertical well]. Production heterogeneity is much greater in the NFRs compared with "conventional" fields **(Fig. 1.1)**. The broad distribution and extended tail of PIF values greater than 20 indicates significantly greater heterogeneity in the NFRs than in the conventional reservoirs.

Various fieldwide, well-specific productivity parameters can be inspected to assess heterogeneity, including productivity index (PI), cumulative production, permeability thickness (kh), flow capacity index (FCI), and absolute open flow (Nelson 2001). **Fig. 1.2** shows the initial potential (IP) distribution from a naturally fractured tight-gas sandstone. It is strongly asymmetric and shows a significant tail of data extending to about 10 times its me-

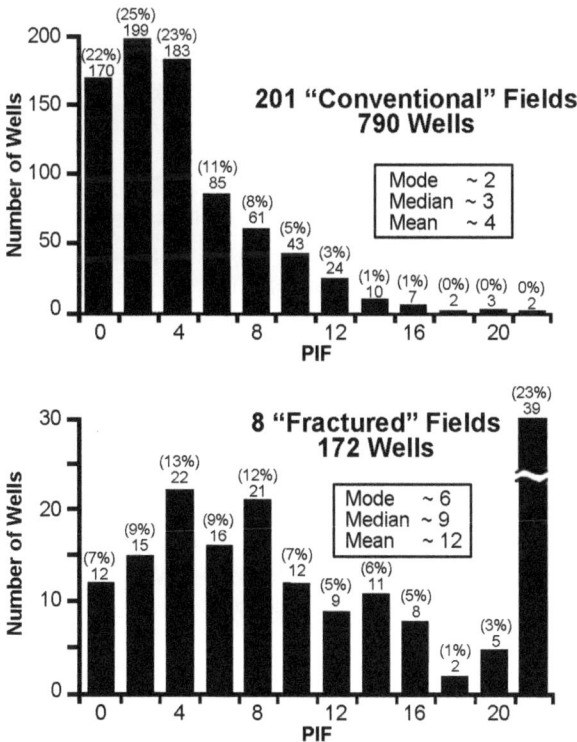

Fig. 1.1—Productivity improvement factor distribution for wells in "conventional" and NFRs (Beliveau 1995).

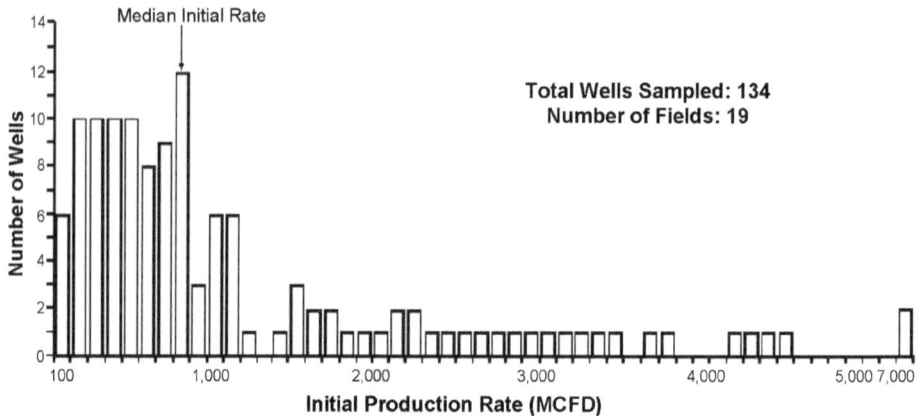

Fig. 1.2—IP distribution of a naturally fractured tight-gas reservoir: the Frontier formation of western Wyoming, from Haas et al. (1989).

dian value. Because of the high variability, it is not uncommon for a few wells in an NFR to engender the financial success of an entire field. An 80/20 rule of thumb is commonly cited for NFR well production: 80% of production is obtained from 20% of the wells in a field. This rule may be a slight exaggeration; nevertheless, it is a good reference point to bear in mind.

Heterogeneity can be used to help recognize the importance of fractures in a reservoir, and production statistics for wells can provide a quick-look method to recognize heterogeneity. A frequency distribution plot of comparable well performance measures gives an indication of heterogeneity. These comparable performance measures include values such as PI, IP, absolute open flow, monthly production, or cumulative production (production values must be used thoughtfully to ensure they are truly comparable from well to well). A strongly skewed distribution (Figs. 1.1b and 1.2) is suggestive of fracture-based heterogeneity, whereas a less asymmetric pattern indicates low-productivity heterogeneity.

It is critically important to examine the spatial distribution of the well productivity whenever a productivity distribution plot is used to judge fracture-induced heterogeneity. This can be done by comparing the production-related measures on a bubble map. If the productivity is variably mixed throughout a field, then fractures are a likely cause of the heterogeneity (**Fig. 1.3**). A well that intersects a large or well-connected fracture will be highly productive, whereas a nearby well may perform poorly if it does not intersect any fractures. However, if the map shows that production is fairly uniform within areas and varies gradually across the field, then the production heterogeneity might be due to lithologic changes that affect the rock matrix. Each of these scenarios can result in a strongly skewed production distribution, but the map can provide evidence as to whether fractures are the probable cause.

Another powerful tool for recognizing and assessing the possible flow enhancement due to natural fractures is by comparing well performance *expected* from the matrix with that *observed*. A procedure for accomplishing this is to compute the flow capacity index or FCI, which is the ratio of observed well performance to well performance predicted from matrix properties. This is computed as:

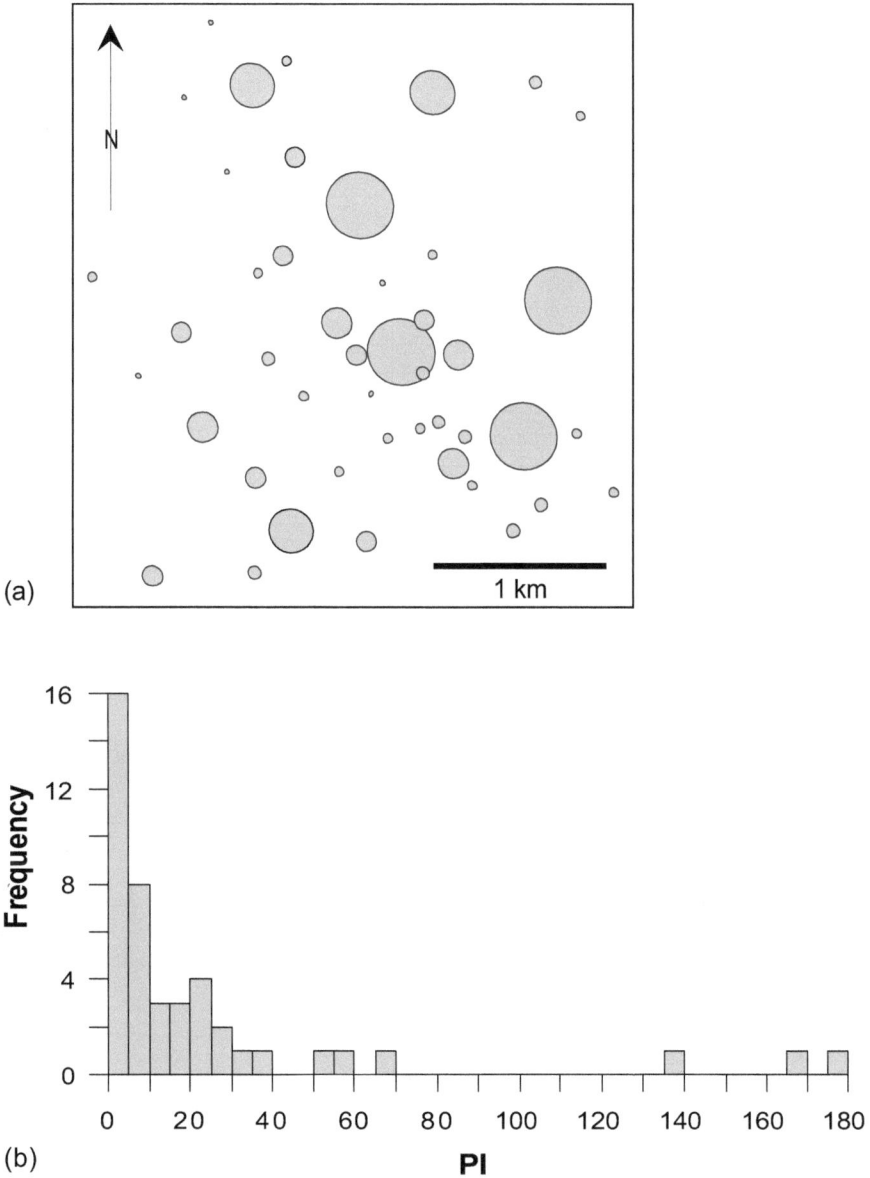

Fig. 1.3—(a) Bubble map showing PI distribution of an NFR; bubbles are centered on well loca-tions, and their size (area) is proportional to well PI. (b) Frequency distribution plot of PI for the wells in bubble map (a). Notice the erratic distribution organization of highly and poorly pro-ductive wells in the bubble map, which is typical of an NFR.

$$FCI = \frac{kh_{well}}{kh_{matrix}}, \dots\dots\dots\dots\dots\dots\dots\dots\dots\dots\dots\dots\dots\dots\dots (1.1)$$

where kh_{well} is measured from a well test, whereas kh_{matrix} is computed based on the matrix permeabilities determined for the rocks across the zone of well completion. If the value of FCI is near unity, then the well is performing approximately as predicted by the matrix. If the value is much greater than unity then the well is "overperforming," and the explanation may be effective fractures.

The concept seems to have been suggested originally by Reiss (1980) and is referred to by some authors as FPI for "fracture productivity index." Other reservoir heterogeneities— such as thin high-permeability beds or connected vug networks—can enhance the well kh; therefore, FCI is a better term because it acknowledges the potential for nonfracture wellbore-flow enhancements. Semantics notwithstanding, FCI can be a good indicator of fracture-controlled productivity enhancement.

The histogram of **Fig. 1.4** shows the distribution of FCI for several wells in a sandstone reservoir. Seven of the wells have low to moderate FCI values (less than two) that are in the range of productivity variation for conventional reservoirs. But the remaining six wells show FCI values greater than four. This is common in NFRs and is a sign of heterogeneity. Indeed, the range of FCI values for this example is low compared to many NFRs. Furthermore, these are vertical wells, and the fractures are also vertical, thus minimizing the probability of fracture intersection. The seven low-FCI wells may not intersect any effective fractures; their observed kh_{well} is close in value to kh_{matrix}.

Another way of plotting FCI data is to cross-plot kh_{matrix} vs. kh_{well} (**Fig. 1.5**). This plot shows the dominant behavior of wells fieldwide, and it gives a sense of the variability in these data. If wells perform approximately according to matrix-based prediction, they

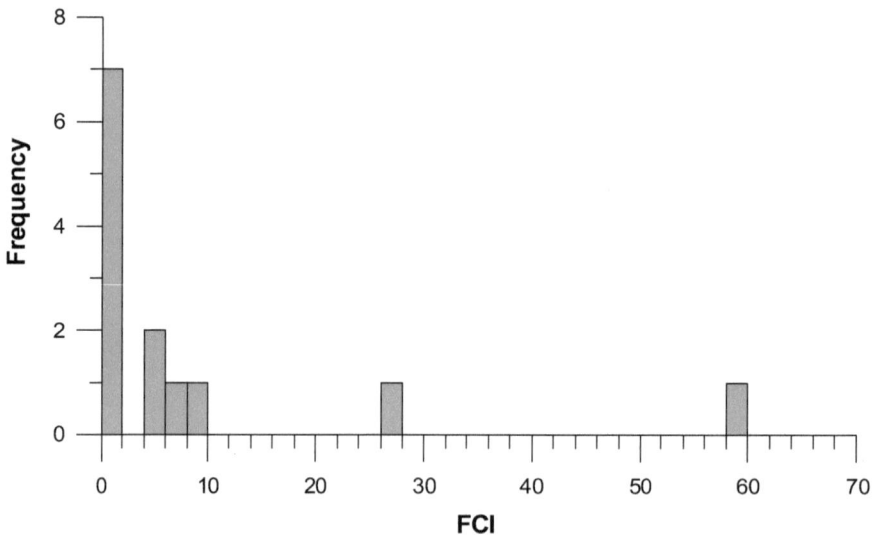

Fig. 1.4—Distribution histogram of FCI in wells in a gas field producing from a naturally fractured sandstone reservoir.

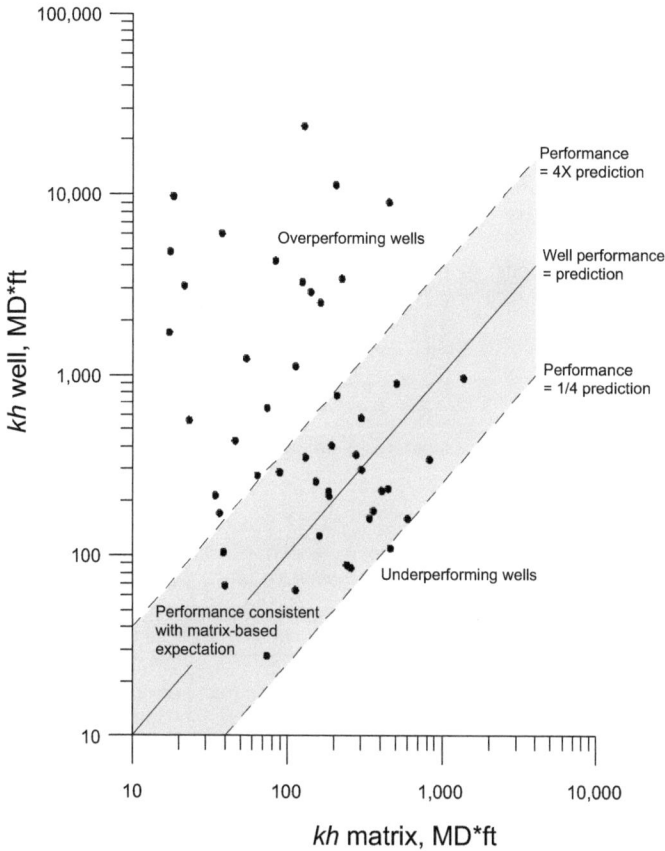

Fig. 1.5—A plot of kh_{well} vs. kh_{matrix} shows the degree to which the reservoir flow behavior departs from that predicted based on matrix properties. Each point represents a well in a carbonate reservoir. The data show that many wells perform significantly better than predicted, which can be indicative of reservoir fractures. The width of the zone of "matrix-based expectation" is a factor of four here, but may be defined according to local experience.

should cluster around a line with a slope of 1. The data in Fig. 1.5 show that a majority of wells significantly outperform predictions, which is due to fracture enhancement in this field.

The height of the producing interval in an NFR may be less than the total open interval in a well and less than the thickness of the defined reservoir. Assessing the correct value of the productive height (*h*) can be challenging; in a low-matrix-permeability NFR, the production may be almost exclusively from the fractures. After adjusting for the height of the actual effective producing interval, the value of FCI can become quite large. A downhole flowmeter or production logging tool (PLT) log is useful to help determine the producing interval height.

Various other data can be used to help recognize whether a reservoir contains significant fractures (**Table 1.2**). These data are best used in combination, and with production information as described above. The table lists specific observations that can indicate reservoir fractures, the explanation of the phenomenon, and the precision to which the observation

TABLE 1.2—INDICATIONS OF FRACTURES IN A RESERVOIR*

Observation	Possible Explanation	Precision
Production/flow-related evidence		
Isolated intervals of high productivity on PLT that do not correspond with good reservoir rock	Fractures enhance permeability	Specific depth in well
PI of well exceeds expectations	Fractures enhance permeability	Reservoir
kh_{well}/kh_{core} >>1	Fractures enhance permeability (core analyses should exclude samples with fractures)	Well or reservoir
Low porosity, low resistivity	Low fracture porosity, but good conductivity due to fractures filled by drilling fluid	Specific depth in well
Significant unexpected gas show or "kick," especially in low-permeability rock	Light hydrocarbons in fracture	Specific depth in well
Dual-ϕ behavior in well test	Fracture drainage, followed by matrix drainage; the absence of this effect is not evidence against the presence of fractures	Reservoir
Highly productive well shows rapid decline in productivity	Fractures intersected by well are of limited extent, and drain quickly, so system reduces to drainage from matrix	Reservoir
Rapid rise in gas/oil ratio to water/oil ratio (GOR/WOR), then stabilization at a level >> solution GOR	Dual system behavior; fractures drain first, then gas and water enter, then matrix feeds fractures, and a new steady state is achieved	Reservoir
Water or gas coning	High vertical conductivity of fractures allows fluids to be rapidly drawn across stratigraphic layers to low-pressure area near producing well	Reservoir, specific depth
High well-to-well productivity variability	Fracture intersections can result in great variation in well performance	Reservoir
Well-to-well hydraulic communication	Movement of fluids between wells along aligned fractures	Reservoir
Rapid breakthrough of injected fluids along a consistent trend	Movement of fluids between wells along aligned fractures	Reservoir
High vertical connectivity	Fractures cross-cut low-k layers	Reservoir
Strong anisotropy in multiwell pulse test	Rapid hydraulic conductivity along oriented fracture set	Reservoir

*Detailed well-log evidence is discussed in Chapter 3.

may be applicable. Some data are specific to a certain depth in a well, such as a fracture seen in core, whereas other data are more broadly based—such as a multiwell pulse test.

The information in Table 1.2 is presented as a qualitative checklist to help in recognition of an NFR, but many of these points are discussed at greater length in Chapter 3, in a context more useful for quantitative reservoir characterization. The table is organized according to the source of information. None of these observations should be used in isolation to resolve the fractured nature of a reservoir. The evidence used to judge whether a reservoir is naturally fractured is circumstantial; hence, more evidence is better.

TABLE 1.2—INDICATIONS OF FRACTURES IN A RESERVOIR (continued)		
Observation	Possible Explanation	Precision
Drilling evidence		
Drilling breaks	Rapid drilling progress along fractures, especially through intervals where rapid drilling is unexpected	Specific depth in well
Bit drop	Open fracture, usually enhanced by solution enlargement to create a large open cavity	Specific depth in well
Bit chatter	Bit-fracture interaction	Specific depth in well
Abrupt loss of circulating fluids	Mud drained into extensive open fracture system. Losses can damage reservoir	Specific depth in well
Geological evidence		
Open natural fractures evident in core or image logs	Fractures in the reservoir are open	Specific depth in well
Spikes in Stoneley wave reflectivity log	Fractures are open in the reservoir	Specific depth in well
Hard rock in the reservoir	Brittle rock has a tendency to respond to strain by creation of open fractures	Reservoir
Same reservoir productive from fractures elsewhere	Subsurface fractures are a characteristic of this formation	Reservoir
Geophysical indications of fractures		
Shear wave polarization in seismic, vertical seismic profile (VSP), full-wave sonic	Birefringence due to open fractures. Effect may also be caused by stress-induced birefringence (fractures not necessary)	Portion of reservoir
Shear and P-wave velocity anisotropy in VSP or surface seismic	Seismic wave velocity varies as a function of raypath orientation relative to fracture set orientation	Portion of reservoir
Anisotropic seismic attenuation	Seismic attenuation varies as a function of raypath orientation relative to fracture set orientation	Portion of reservoir

Due to the heterogeneous nature of NFRs, the evidence will be uneven. For example, one well may lose circulation returns, and then the following two wells may drill through without significant losses. A heterogeneous response of the reservoir to the indicators in Table 1.2 is yet another indication that the reservoir is fractured.

1.3 Example From Spraberry

The discovery and development of the Spraberry trend area in west Texas beginning about 1950 provides a good example of the diagnostics that eventually led to classification of the reservoir as fractured. The chronological events described are typical of early recognition of the effect of fractures on reservoir performance.

1.3.1 Rapid Decline in Primary Production. The Spraberry trend area in west Texas was discovered in 1949. By 1951, it was understood this vast, deep submarine fan was extensive in area and contained approximately 10 billion bbl of oil. The effect of fractures was evident even in the first few wells. Although Spraberry wells initially flowed at high rates, production declined rapidly because, in spite of the fact that the reservoir was highly fractured, the matrix permeability was low and could not respond fast enough to support the high rate of production from the fractures.

Unusually rapid pressure decline during primary depletion was observed in many Spraberry wells. Eventually, production from the fractures was depleted to the point where the matrix rock began to contribute a larger fraction of the production. The fractures resulted in high IPs, and once the fracture porosity had been produced, the wells stabilized at a low but steady rate as the matrix fed the fractures. If the matrix permeability is low—as in the Spraberry case—then some wells can sustain low production rates for many years with very little decline.

Primary depletion of reservoir pressure is observed in **Fig. 1.6**. The initial pressure was near 2,300 psi. Rapid decline of pressure led to production below the bubblepoint over a

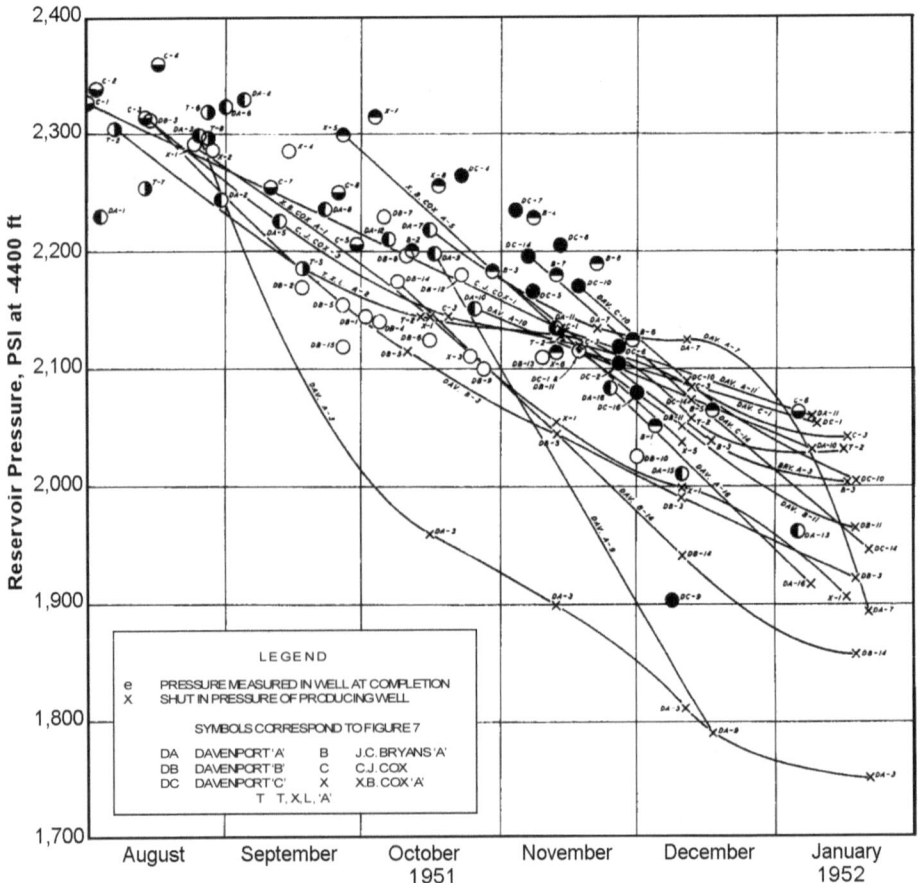

Fig. 1.6—Rapid decline in producing and shut-in pressure in many wells during primary depletion in Spraberry indicates high connectivity of the fracture system (from Elkins 1953).

wide area. An interesting observation shown in Fig. 1.6 is that all the wells, regardless of whether producing or shut-in, declined at the same rate. This implied widespread connectivity of the fracture system. Elkins (1953) thus proposed that one well could effectively drain 160 acres. This concept was to be challenged in the 1990s as small companies with low drilling and completion costs infill drilled, targeting the matrix rock. Although the IP is modest and rapidly declines to 10–20 barrels of oil per day, operators are still able to produce Spraberry economically on 40 acres, owing to the long life of the wells as the matrix porosity is gradually depleted.

1.3.2 Large Differences Exist Between Well-Test Permeability and Core Permeability. The first wells drilled in the naturally fractured Spraberry trend area flowed at up to 1,000 BOPD (Christie and Blackwood 1952). Core samples obtained from some of the first wells showed that the siltstone pay zones have very low matrix permeability (no greater than 1 md). Well tests indicated reservoir permeability on the order of 100 to 200 md (FCI = 100 to 200) (Dyes and Johnston 1953). The large discrepancy between laboratory-measured permeability on core samples and well-test-derived permeability provided the first indication that natural fractures dominated well performance. Even though fracture porosity in Spraberry is less than 0.1%, drainage of well-connected fractures over large distances was clearly the reason for the high IPs in the first wells.

1.3.3 Directional Permeability. Fractures occur in extensive, consistent parallel groups or sets, and their orientation determines the direction of flow anisotropy. As early as 1951, permeability anisotropy was recognized in Spraberry (Wilkinson 1953). A NE/SW preferential flow orientation is evident from IPs mapped across several square miles in the Tex Harvey Spraberry Unit **(Fig. 1.7).**

1.3.4 Rapid Increase in Gas/Oil Ratio (GOR). Rapid decline in reservoir pressure in the Spraberry trend hastened the increase in GOR. GORs increased from solution gas levels of 1,000 scf/STB (standard cubic feet per stock tank barrel) up to 20,000 scf/STB within the first year of development. Infill drilling exacerbated this problem, and Elkins (1953) recognized that fracture connectivity caused reduction in reservoir pressure in shut-in wells (as measured by static borehole pressure) at the same rate as pressure decline in nearby producing wells. Pressure measurements in newly completed wells in the Spraberry reservoir also exhibited the same reduced pressure as nearby wells on 160-acre spacing. Pressure communication over large areas was an indication that the fracture network was well connected. **Fig. 1.8** shows that as the pressure drops below the bubblepoint pressure, the GOR increases dramatically from solution GOR of 1,000 scf/STB to over 20,000 scf/STB in some instances.

As the pressure in the fracture system was depleted, solution gas formed near the fracture faces. Because the pressure drawdown was greatest near the fracture face, the gas saturation was highest near the fracture face, thereby reducing the relative permeability to oil. The net result, as shown in **Fig. 1.9**, is that as the fracture network was rapidly depleting, the recovery efficiency was very low—in the 7–8% range.

1.3.5 Directional Fluid Movement. Later evidence verified directional permeability in the Spraberry trend area. A waterflood pilot conducted in the 1950s demonstrated that injected water rapidly moved in a NE/SW direction. Barfield et al. (1959) describe a model

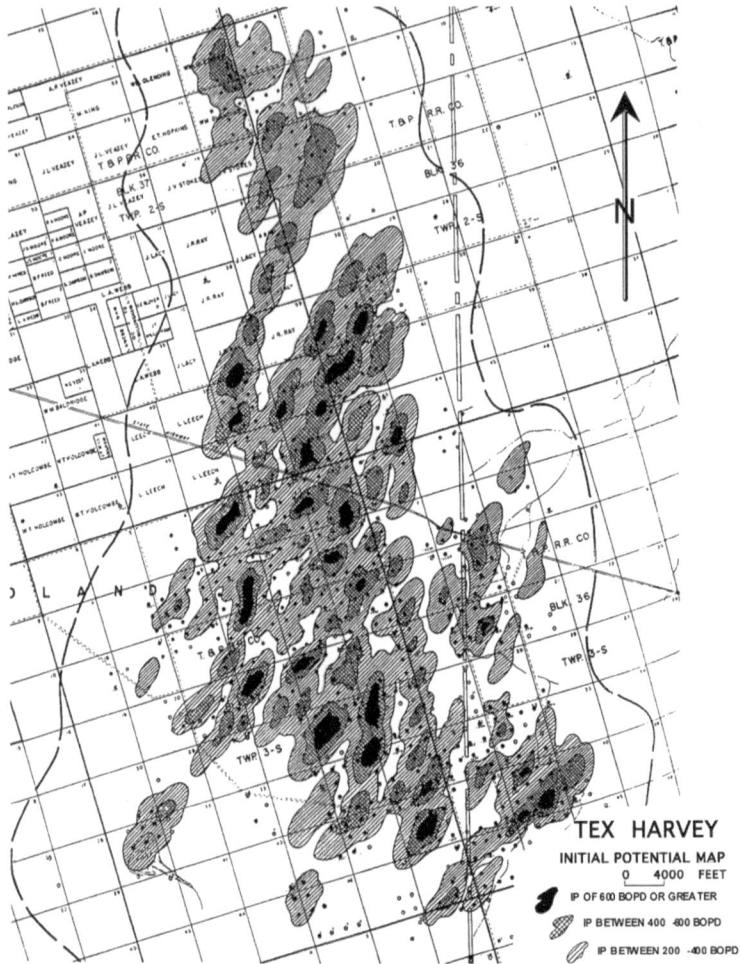

Fig. 1.7—Initial potential contours oriented approximately N30°E in the Tex Harvey Spraberry unit in 1951, showing evidence of directional flow during primary depletion (from Wilkinson 1953; AAPG © 1953; reprinted by permission of the AAPG, whose permission is required for further use).

developed to gauge the movement of injected water. The streamlines in **Fig. 1.10** show significant permeability anisotropy as a consequence of the NE/SW direction of the natural fracture system. Ratios as large as 150:1 on-trend to off-trend permeability were recorded in this waterflood pilot.

Comparing the IP anisotropy contours of Fig. 1.7 with the water-breakthrough trend of Fig. 1.10 shows a 20° difference in orientation. This may represent different fracture trends at these different sites. Alternatively, this difference may be a consequence of differences of interpretation from these separate data sources. The IP anisotropy interpretation of Fig. 1.7 is particularly subjective; hence, a precise anisotropy direction is not expected. The uncertainty of these dynamically defined interpretive trends could be reduced by integrating them with static geologic data on fracture orientation from core or image logs.

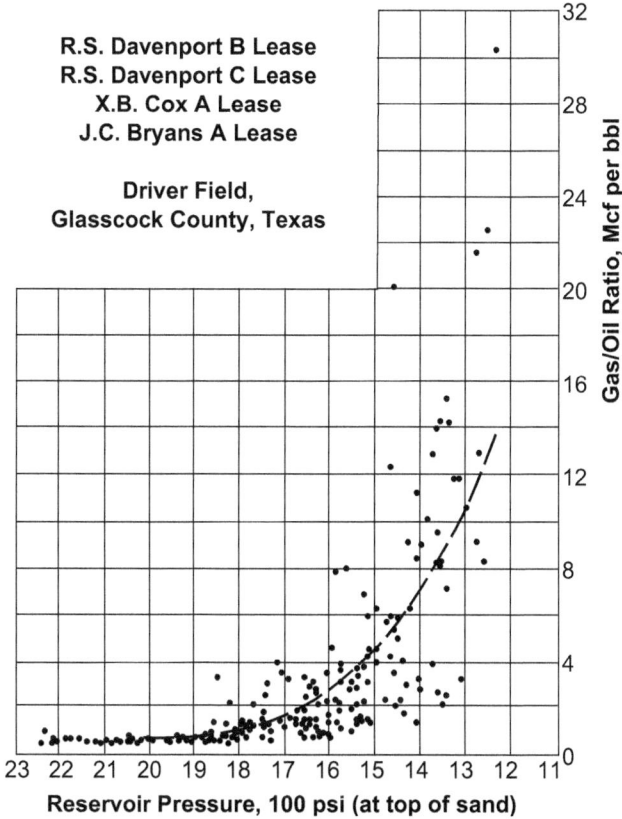

R.S. Davenport B Lease
R.S. Davenport C Lease
X.B. Cox A Lease
J.C. Bryans A Lease

Driver Field,
Glasscock County, Texas

Gas/Oil Ratio, Mcf per bbl

Reservoir Pressure, 100 psi (at top of sand)

Fig. 1.8—Rapid increase in GOR as the fracture system was produced, reducing the reservoir pressure below the bubblepoint pressure of 1,600 psi (from Elkins 1953).

1.4 Understanding NFRs

From the material reviewed in this chapter, it is evident that NFRs exhibit strong anisotropy, heterogeneity, and distinctive production behavior. In subsequent chapters, we begin with an overview of the nature of natural fracture systems, emphasizing those properties significant to fluid flow. Following this is a discussion of methods used to characterize NFRs, from the fine details detected in individual wells to the amalgamation of multiple data sources for fieldwide synthesis. Then, a review of reservoir simulation discusses options available for NFRs. Finally, case histories of several NFRs are reviewed in some detail, to provide a sense of the similarity and diversity of this challenging reservoir class.

The material included in this publication is meant to be a general synopsis of the nature of NFRs, and to provide some practical concepts for evaluating and understanding their productivity characteristics. It is not written for the NFR specialist, nor for the person looking for a comprehensive guidebook on engineering or reservoir geology of NFRs. To understand and work effectively with NFRs draws on both geological and reservoir engineering disciplines. We hope this publication will provide a greater appreciation of the value that can be obtained by uniting our efforts across this discipline divide.

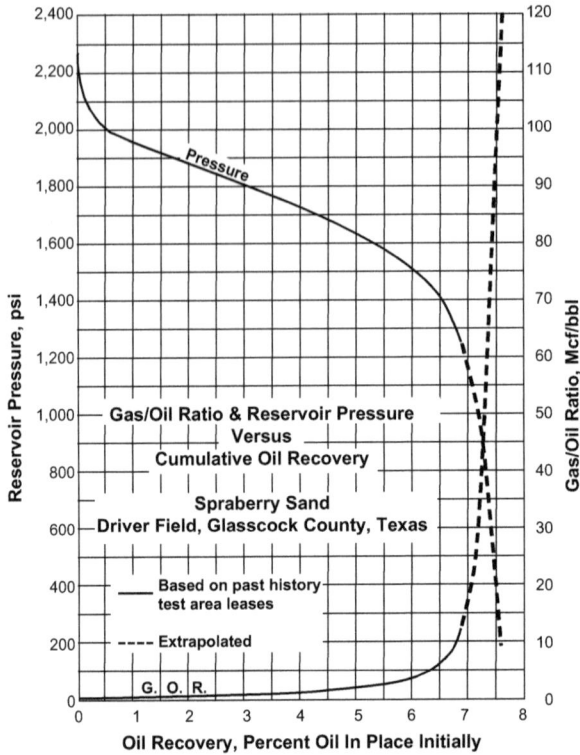

Fig. 1.9—Rapid depletion of Spraberry resulted in high GORs at relatively low recovery factors (from Elkins 1953).

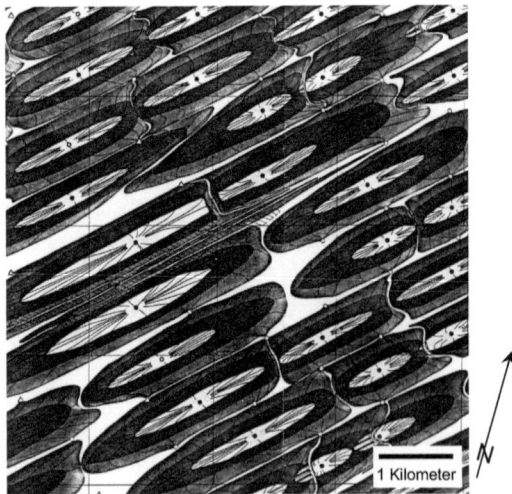

Fig. 1.10—Anisotropic flow of injected water during a waterflood pilot in the 1950s. The permeability ratio is approximately 144:1. The axis of maximum permeability is oriented approximately N50°E (from Barfield et al. 1959).

Chapter 2

Geology of Natural Fracture Systems

2.1 Introduction

The natural fracture systems responsible for NFRs are extremely common in the Earth's crust, and examples may be seen at most outcrops. Fractures have an orderly set of characteristics, particularly in sedimentary rock—our primary interest. This chapter reviews characteristics and occurrence of the most important fracture types encountered in NFRs and discusses the origin of natural fractures. Most illustrations in this chapter come from outcrops, because they provide a more complete view of fractures than can be obtained from the subsurface.

2.2 Fractures

A fracture is a discontinuity or parting in a material caused by brittle failure. "Fracture" is a general term that includes various natural and induced features. **Fig. 2.1** is a conceptual illustration of fractured sandstone.

The two principal naturally occurring fracture types are joints and faults **(Figs. 2.1 and 2.2)**. These are distinct styles of fractures with fundamentally different origins, characteristics, occurrences, and impacts on reservoir fluid flow. Joints are extension fractures; they form in tension or effective tension. Opposite walls of this style of fracture pull directly away from each other during formation **(Fig. 2.3a)**, thus there is no shearing displacement parallel to the fracture surface. Faults are shear fractures, or extension fractures on which later shearing displacement has occurred (Figs. 2.3b through 2.3d). Shearing displacement is the fundamental characteristic that discriminates faults from joints. **Table 2.1** summarizes fracture definitions.

The orientation of both faults and joints is controlled predominantly by the orientation of the Earth's stress field, which varies in direction and magnitude with location. In the environment where most petroleum fields occur—the upper crust—the stress field is anisotropic, and can be resolved into three orthogonal principal components. One of these principal components is nearly vertical; hence, the other components are horizontal (Fig. 2.3). Faults can form with different orientations relative to the Earth's surface, depending on the relative magnitudes of the principal stresses. However, despite the different names and relative orientations of the three main fault types, mechanically they are just shear fractures of different orientation.

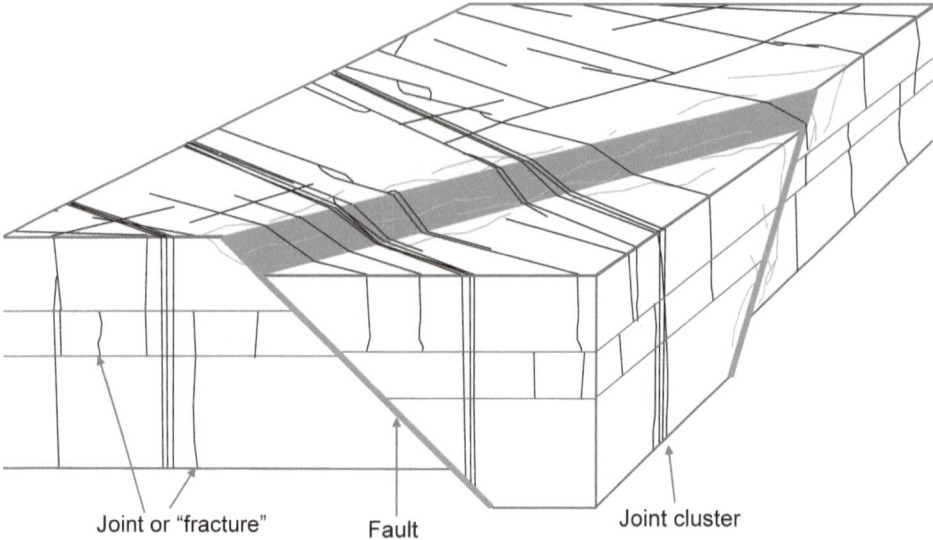

Fig. 2.1—Conceptual view of fractured sandstone showing joints, joint clusters, and a fault.

Fig. 2.2—Well-bedded sandstone and shale strata showing a single reverse fault and many joints, Delaware basin, Texas.

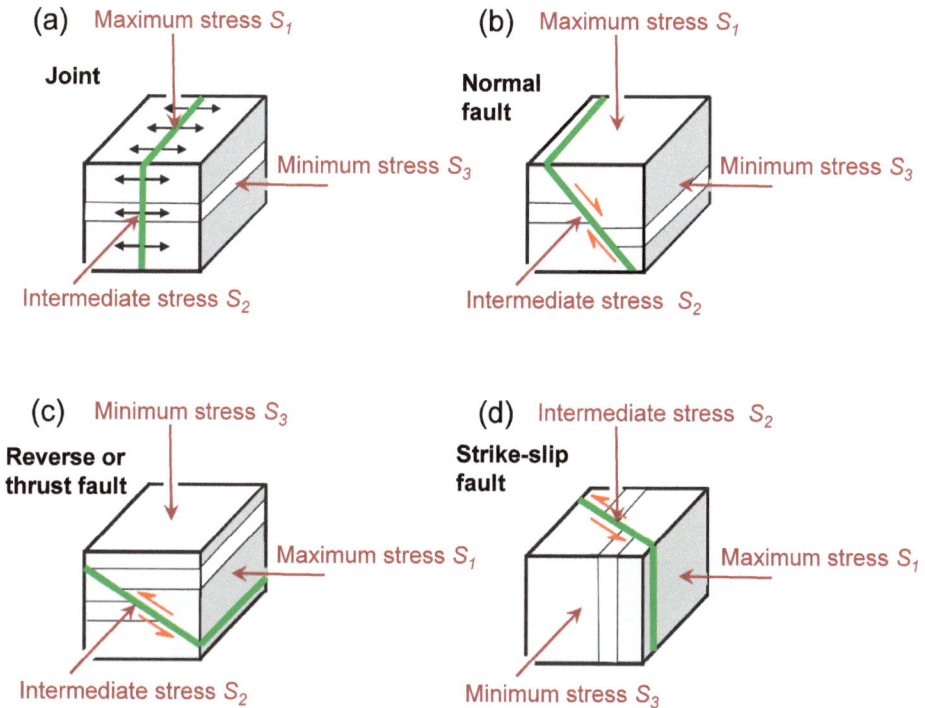

Fig. 2.3—Common fracture styles, their displacements, and their orientations relative to principal stress orientations common in the Earth's upper crust.

2.3 Joints

Joints are natural fractures that show no shearing offset and, thus, are opening-mode or extension fractures (Fig. 2.3a). In bedded rock, joints are typically at high angle to layering **(Figs. 2.1, 2.2, and 2.4)**. Joints commonly terminate at discontinuities in rock or at boundaries between stratified layers. A *joint set* is a group of parallel, spaced joints. Joints virtually always occur in sets. A *joint system* refers to the occurrence of multiple joint sets in an area. Joints are responsible for reservoir-wide enhanced transport of fluids in most NFRs. Suppe (1985) presents an excellent overview of joints.

Joints usually show consistent orientation over large areas—up to regional extent **(Fig. 2.5),** but they can also vary locally across distances on the order of 100 meters. Joints can show consistency in orientation across stratigraphy as they cross different rock types. but they can also change direction abruptly at bedding surfaces.

Most rock outcrops show two or more sets of joints. to form an interconnected network **(Fig. 2.6),** and generally at least two of the sets are at high angle to each other (>70° dihedral angle). However, a single predominant set of joints is more common in the subsurface **(Fig. 2.7).**

Joints in a rock do not all form at the same time, and joint sets do not all form at the same time either. The relative age of joints can sometimes be determined at outcrop, such as

TABLE 2.1—CHARACTERISTICS, DEFINITIONS, AND ORIGINS OF NATURAL FRACTURES OF WIDESPREAD IMPORTANCE TO THE PETROLEUM INDUSTRY

Fracture Term	Definition	Identifying Characteristics, Emphasizing Borehole-Based Evidence
Fracture	Discontinuity caused by brittle failure. All the features defined in this table are fractures.	Distinct discontinuity that post-dates formation of the rock.
Crack	Individual, isolated fracture showing no shearing offset. May be natural or induced.	Isolated (not necessarily part of a set). No shear offset.
Joint	One of a group (set) of naturally occuring spaced, parallel fractures showing no shearing offset. In stratified rock, joints are usually at a high angle to layering. Joint sets are extensive. Multiple joint sets form a "joint system."	No shear offset. At high angle to bedding in sedimentary rocks. With parallel fractures forms a set.
Fault	Naturally occurring fracture along which opposite sides have been displaced parallel to the fracture surface.	Shear offset. Fault-induced deformation adjacent to fault.
Deformation band	Planar, failure-induced discontinuities containing disaggregated or broken particles of host rock. They form in porous, granular strata. They are usually small-offset faults, but they have been seen with no detectable offset. They often occur in parallel groups or sets, and are commonly conjugate.	Shear offset (typically), braided appearance. Occur in groups with parallel, and frequently conjugate, faults. Occur in porous, granular strata.
Vein	Fracture filled by precipitated mineralization.	Mineral-filled fracture.

*Nomenclature presented here derives from an unpublished 1998 manuscript, "Fractured Jargon," by W. Narr, T. Engelder, A. Lacazette, and E. Willemse.

Fig. 2.4—Joints in a well-bedded sandstone. Joints are perpendicular to bedding, and they commonly terminate at mechanical bed boundaries. The joints are parallel, and their spacing is greater in thicker beds.

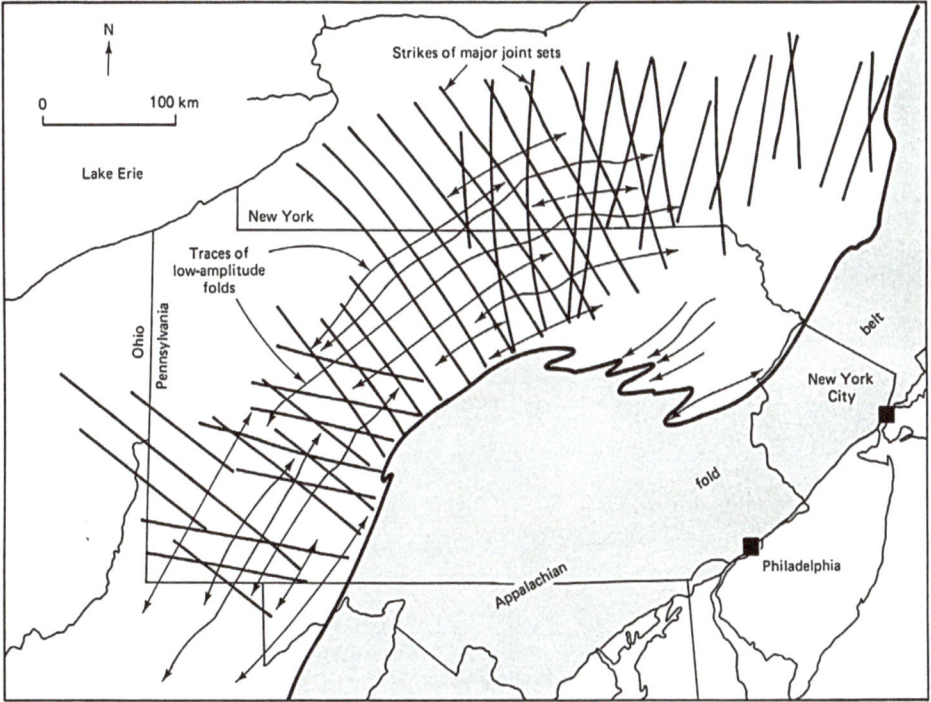

Fig. 2.5—Dominant joint trends in siltstone and shale of the Appalachian Plateau of Pennsylvania and New York (from Suppe 1985, after Nickelsen and Hough 1967).

Fig. 2.6—Vertical aerial photograph of three joint sets in flat-lying Windgate sandstone, Arches National Park, Utah. Joint set A is the oldest, followed by B, then C (photo courtesy R. Dyer).

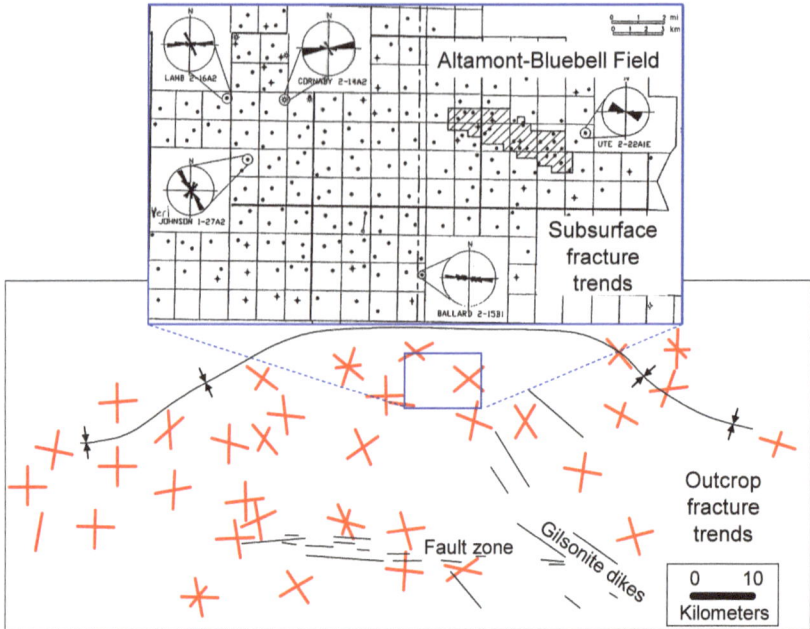

Fig. 2.7—Joint trends from surface outcrops throughout the central Uinta basin (bars) compared with orientation-frequency rose diagrams of joints measured in the reservoir of Altamont-Bluebell oil field, 3000 to 4000 m depth (Montgomery and Morgan 1998; Narr and Currie 1982; AAPG © 1982, 1998; reprinted by permission of the AAPG, whose permission is required for further use).

Fig. 2.6. The joints parallel to set A formed earlier than set B. This is evident because the set B joints terminate against set A joints; therefore, set A was already present before set B formed. Joints of set C terminate against both set A and set B; therefore, they are the last set to form. Joints in set A are long and linear, set B joints are linear, but set C joints are short and crooked. This crookedness suggests that set C formed when the differential stress field on the rock was considerably lower than when sets A and B formed—probably during the final stages of erosional unloading.

Generally, if multiple fracture sets are present in the subsurface, their relative age cannot be determined from their geometry alone. Sets A and B in Fig. 2.6 are separated by a small dihedral angle. These joint sets clearly formed at different times, so they are not conjugate fractures. Conjugate geometry is a characteristic of shear fractures (faults), so we shouldn't expect joints to form conjugate sets.

2.3.1 Fracture Geometry. Aperture, height, length, and orientation are the basic geometry elements of individual joints. Aperture is the gap between the walls of a fracture, which may be open and contain some quantity of mineralization. The aperture of a fracture is important to its ability to transport fluid. Trustworthy aperture data is difficult to collect from outcrops unless fractures are mineralized, because open apertures are commonly weathered.

Height is the fracture dimension perpendicular to bedding. Many stratigraphic sequences show a large number of short joints and relatively few tall joints. For example, the thin layers toward the bottom of the outcrop in Fig. 2.4 have many short joints, but only a few tall joints cross the entire outcrop. Joints often terminate at bed boundaries, where mechanical changes in rock property cause fracture propagation to stop. Fracture height can affect gravity-induced fluid drainage from a reservoir; tall fractures are more effective at draining units than short fractures. **Fig. 2.8** shows fractures and their height distribution in relatively massive sandstone of the Weber formation of central Wyoming.

Joint length is especially important to effective permeability of a reservoir. Long fractures can quickly transport fluid across long distances. Fractures in sedimentary rocks are typically much longer—10 to 100 times longer—than their height. Fracture length is a poorly understood aspect of joints because exceptionally extensive outcrops are needed to

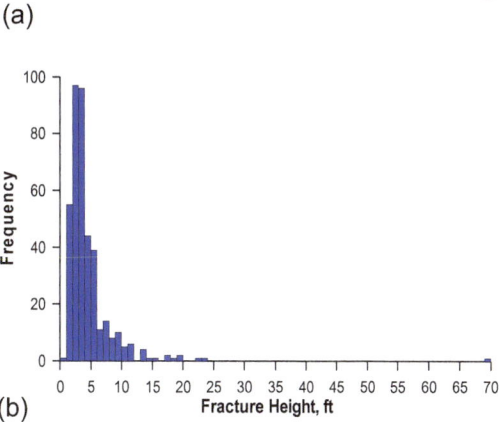

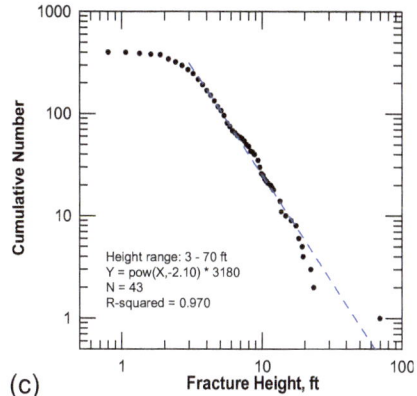

Fig. 2.8—(a) Massive eolian sandstone of the Weber Formation, Whiskey Gap, Wyoming, contains fractures having a variety of heights. Bedding is approximately horizontal in this view; joints are approximately vertical. (b) Height distribution of joints in (a). (c) Cumulative frequency vs. fracture height of data in (a).

evaluate length; hence, relatively little observational work has been done on the topic. For example, despite its large, vegetation-free extent, many of the longest fractures in the aerial photograph of **Fig. 2.9a** are not visible from tip to tip because they extend beyond the edge of exposed rock outcrop. This is particularly frustrating because the longest fractures are of greatest interest when trying to understand potential fluid-flow pathways. As with height, the length distribution of joints is strongly skewed such that there are many small joints for every long one. Fracture length distributions, as well as height distributions, commonly show exponential or power-law distribution forms (Figs. 2.8c and 2.9b) (Gillespie et al. 1993; Marrett 1997). The length distribution in Fig. 2.9b shows two distinct linear segments, which may indicate a difference in mechanics of fracture development for shorter joints and longer joints.

Joints at the short and long ends of the distribution curves in Figs. 2.8c and 2.9b depart from the well-defined linear central region. This is a consequence of nonrepresentative sampling near the tails of the distribution. Joints smaller than a minimum size are not readily visible on photographs or in outcrops, so small fractures are undersampled. Likewise, very tall and very long joints tend to extend out of the area of exposed rock, so the tallest and longest joints are incompletely sampled.

2.3.2 Joint Spacing. The distance between parallel joints is their spacing; spacing affects the effective drainage of the matrix and the effective permeability of the rock. Fracture density for a single set of joints is the reciprocal of the fracture spacing. Fracture density is particularly useful for purposes of NFR characterization and is discussed further in Chapter 3.

Each joint set in a rock typically exhibits a distinct spacing character, as is evident in Figs. 2.6 and 2.9a. Reasons for this may be that joints form at different times, hence under the influence of different stress conditions, and because early-formed sets influence later-formed joints (Gross 1995; Bai and Pollard 2000). Within a single set, the spacing distribution of joints is commonly lognormal, but can range from exponential to normal (Narr and Suppe 1991; Rives et al. 1992; Gillespie et al. 1993).

In rocks with well-defined mechanical layering, joint spacing is controlled strongly by mechanical layer thickness (Fig. 2.4) (Narr and Suppe 1991). Within a given rock type at an outcrop, the bed thickness to joint spacing ratio tends to be constant across a limited range of bed thickness—perhaps one order of magnitude (**Fig. 2.10**).

Joints often cluster into closely-spaced groups separated by relatively unfractured rock (**Fig. 2.11**). Fracture clusters tend to be both long and tall; hence, they can be efficient "superhighways" for fluid flow.

Overall joint spacing is not uniform. This has potentially significant impact for a simulation model running a conventional dual-porosity formulation. The spacing used in a dual-porosity simulation model is uniform within a single cell, and it is often specified as uniform over a large portion of a reservoir. However, the variable spacing of fractures seen in nature suggests that such a model is a highly idealized representation of actual matrix-block geometry. This could have a significant impact on reservoir recovery and performance predictions.

2.3.3 Surface Markings. Features on the surfaces of fractures can give indications of the origin and subsequent history of fractures. These markings are also useful because they can be used to distinguish natural from induced fractures in core and to distinguish joints from faults.

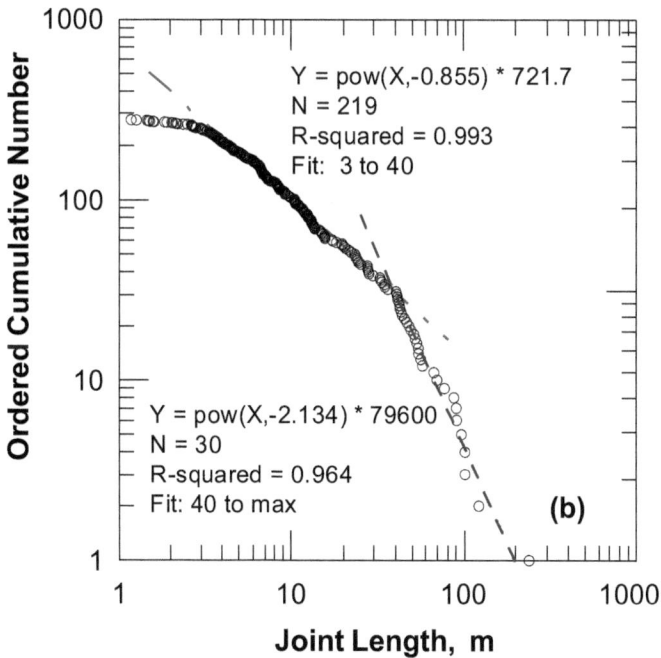

Fig. 2.9—(a) Slightly oblique aerial photograph of Weber formation sandstone, Skull Creek anticline, Colorado. (b) Length distribution of joints that comprise ENE-striking set.

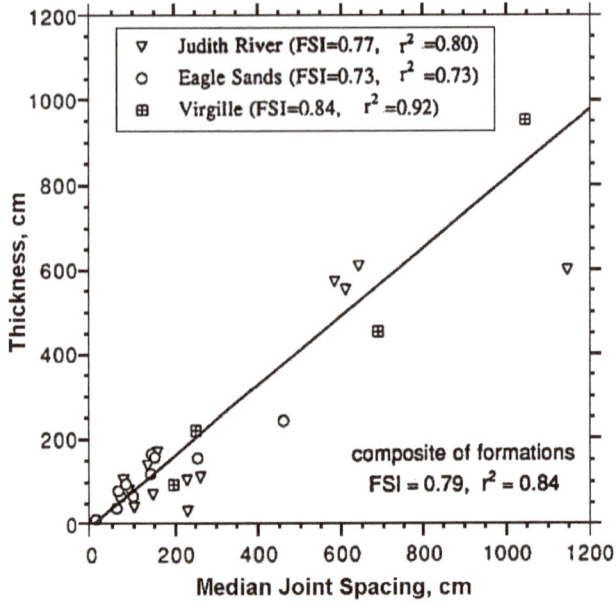

Fig. 2.10—Median bed thickness to median joint spacing shows a linear correlation in the Judith Creek Sandstone, Elk basin, Wyoming (Engelder et al. 1997).

Fig. 2.11—Cluster of vertical joints in thickly bedded Cretaceous carbonate rocks, near Monterrey, Mexico.

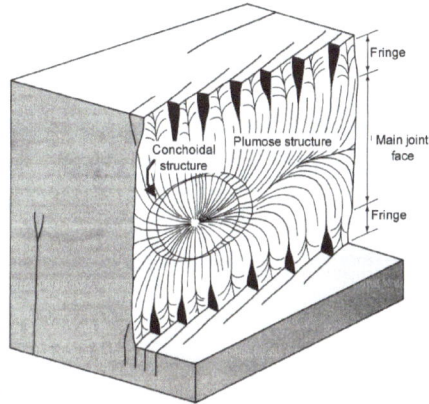

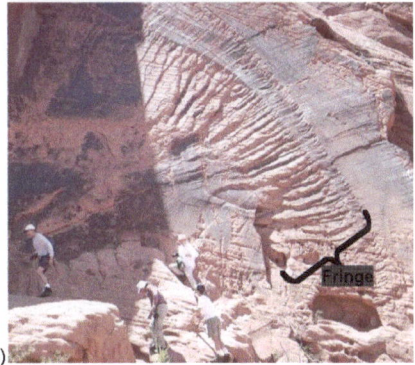

Fig. 2.12—(a) Schematic illustration of a joint showing common surface morphological markings, from Suppe (1985). (b) Plumose structure on siltstone. (c) Concoidal structure on joint surface in sandstone. (d) Fringe zone on joint surface in sandstone. Photos (c) and (d) are the Aztec formation, Valley of Fire State Park, Nevada.

A group of related fracture-surface features form on extension fractures **(Fig. 2.12a)**. Hackle marks (or plumose structures) are subtle features that track the progressive propagation of a fracture, which initiates at a point (sometimes a visible flaw) and commonly shows a pattern similar to an ostrich plume (Fig.12.2b). The propagating fracture can develop concoidal structure—concentric bands perpendicular to the propagation trajectory (Fig. 2.12c). The outer margin of a fracture can subdivide into an array of *en echelon* surfaces as the fracture approaches a mechanical boundary. This region is referred to as "the fringe" (Fig. 2.12d).

Genetically distinctive surface features are also present on shear fractures. These include scratches showing the relative movement direction of fracture surface, stair steps, and secondary fractures induced by shearing on the primary fault.

2.3.4 Diagenetic Alteration of Fractures. Open fractures are efficient pathways for fluid migration, so it is not surprising that minerals deposit on the fracture walls and can be a source of information about the chemical evolution of the rock. A fracture that is completely filled by secondary mineralization is properly called a "vein." The most common fracture-fill minerals in reservoir rocks are calcite, dolomite, and quartz. Mineral growth textures within a fracture can record its displacement (kinematic) history **(Fig. 2.13)**.

Fluids moving through fractures can react with the rock, corroding fracture walls and thereby enlarging their apertures. Carbonate rock—particularly limestone—is especially susceptible to solution enlargement due to the solubility of calcite and other carbonate minerals in acid (James and Choquette 1988). Near-surface acidic water can corrode a carbonate rock beginning shortly after deposition (Mylorie and Carew 1995). Or deep, sulfide-rich fluids associated with hydrocarbon migration can enlarge fractures (Hill 1999). Most cave maps show a rectilinear arrangement of passages, clearly indicating their fracture-system ancestry **(Fig. 2.14)**. Solution enlargement of fractures is common in carbonate reservoirs and can lead to bit drops and massive lost circulation during drilling.

2.3.5 Basement Rock. Geological basement—the metamorphic or igneous rock beneath the sedimentary cover—usually has no effective matrix porosity or permeability; thus, it is not a typical reservoir. But, in rare cases, it has proven to be producible thanks to fractures that provide essential permeability and porosity. Examples include Bach Ho and Rang Dong fields in Vietnam (Areshev et al. 1992), and Mara and La Paz fields in Venezuela (Nelson et al. 2001).

Basement rock tends not to be as reliably layered as sedimentary rock; consequently, some characteristics of joints differ from those in stratified rock due to the absence of me-

Fig. 2.13—This fracture surface has been mineralized by calcite, which bridged across the aperture during shearing displacement. The lineations in mineral growth, and their stair-step edges, show that this is a fault and that the overlying block moved down and to the right. Width of area in photograph is 3 cm.

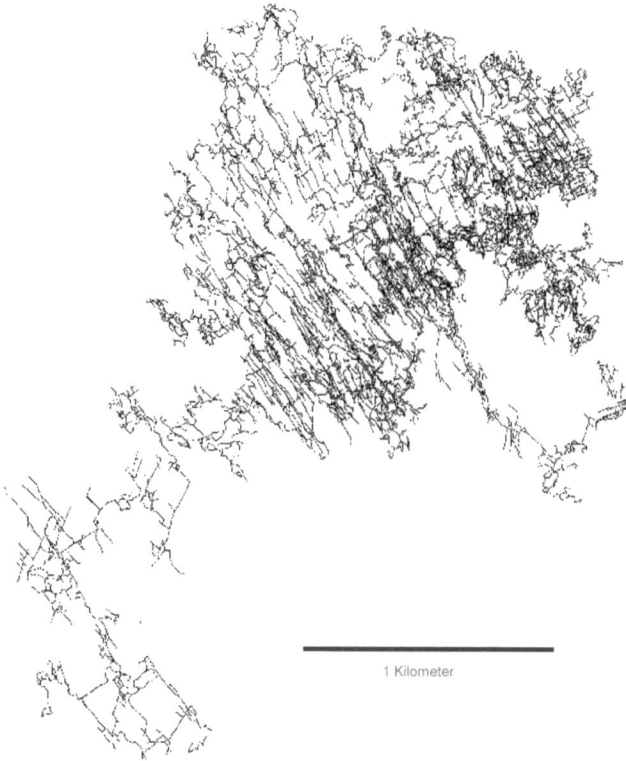

1 Kilometer

Fig. 2.14—Map of Wind Cave, formed in Mississippian limestone in the Black Hills, South Dakota (National Park Service 2004).

chanical layering. The surface between the basement and the sedimentary cover is nearly always an unconformity, which was once exposed at or near the Earth's surface. Joints commonly develop in rock exposed to the surface, and chemical weathering and small-scale movements of the joints at the surface can lead to their enlargement.

In some basement rock, especially intrusive igneous rock, extensive exfoliation (sheeting) joints can develop parallel to the unconformity surface, similar to the layers on an onion. Exfoliation joints are related to erosional unloading and thermal-cooling-induced stress (**Fig. 2.15**).

2.4 Faults

Faults can enhance the flow of fluids through rock, or they can act as barriers to fluid flow, depending on their openness and on the composition and texture of material within the fault zone. Faults are of great significance for their role in compartmentalizing reservoirs, which is the focus of the specialized discipline of fault-seal analysis. Discussion of fault-seal analysis is beyond the scope of this publication.

Faults are fractures along which there has been a shearing displacement (Figs. 2.1 through 2.3). They can have various orientations depending on their tectonic style (Fig. 2.3), but bedding has less influence on their orientation than it does on joints. Faults range

Fig. 2.15—Sheeting joints (exfoliation) of relatively homogeneous granitic rock in Rocky Mountain National Park, Colorado. The rock fractures into layers like an onion.

in size from very small to very large, and the size (surface area) of a fault relates to its displacement [larger size corresponds to more displacement (Scholz et al. 1993)]. The shape of faults commonly ranges from planar to listric (spoon-shaped) to stair-stepped.

Faults can occur in isolation (Figs. 2.1 and 2.2) or in spatially clustered groups. When multiple faults are present, they tend to be either parallel or in conjugate sets (**Fig. 2.16**).

2.4.1 Fluid Flow. Faults can enhance fluid flow locally, but they generally do not enhance reservoirwide permeability. This is a result of both the mode of occurrence of faults and their mode of origin. A fault that is open for fluid flow can have a large surface area, and it can intersect a large number of joints (Fig. 2.1), but a system of hydraulically conductive faults is rarely pervasive through a rock in the sense that joints are. However, if a well should intersect a hydraulically conductive fault, it can have a large impact on well performance. Additionally, faults are apt to be easier targets for wells because they tend not to be vertical, and they may be visible on seismic data.

As shearing slip accumulates on a fault, rock material from both sides can be caught between the fault's walls (Fig. 2.2). The hydraulic-flow behavior of the fault can depend on the composition and physical texture of this material. The two sides of the fault are in contact as the fault slips, thus physically grinding the material between the fault planes, creating "gouge," a material with lower permeability than the host rock. Gouge forms in rocks with a high proportion of strong mineral grains like sandstone. If clay-rich rocks are dragged into the fault zone, the clay minerals form "smear"; this can be a very effective barrier to fluid flow.

The juxtaposition of rocks on one side of a fault relative to the rocks on the opposite side can affect how well fluid moves across a fault. For example, if permeable sandstone is juxtaposed against a mudstone, there is little likelihood of flow across the fault. If permeable rocks are in juxtaposition contact across a fault, there is potential for cross-fault transport. Juxtaposition maps of faults are used to analyze the distribution of strata abutting across a fault. A fault can be regarded as a zone consisting of a central core surrounded by a damage zone where deformation decreases in intensity with distance from the fault. The damage zone is usually localized close to the fault (up to a few meters distance).

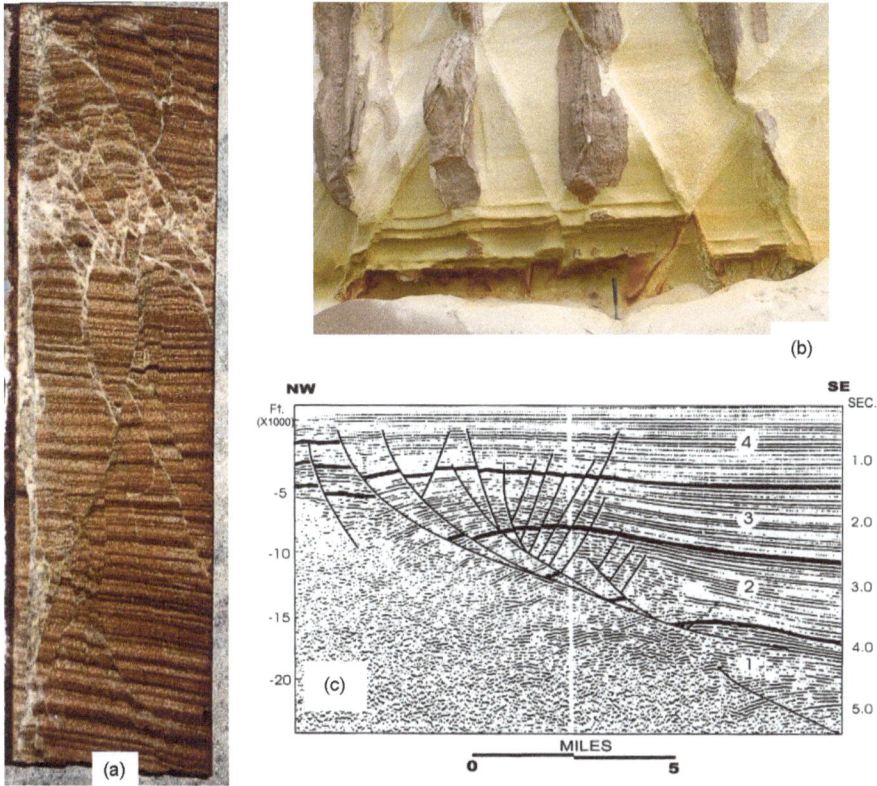

Fig. 2.16—(a) Minor faults in core show conjugate geometry (core diameter 10 cm). (b) Conjugate faults (deformation bands) in Miocene sandstone, San Onofre, California. Dark patches in upper portion of photo are surface weathering of outcrop (pictured area is 4 m wide). (c) Seismic section showing conjugate faults above a major listric normal fault, Gulf of Mexico, Texas (White et al. 1986). Reprinted from White, N.J., Jackson, J.A., and McKenzie, P.P., The Relationship Between the Geometry of Normal Faults and That of the Sedimentary Layers in their Hanging Walls, *J. of Structural Geology*, V.8, 879–909, © 1986, with permission from Elsevier.

2.4.2 Deformation Bands. These small-offset faults can be isolated or distributed over large areas (several square kilometers) (Fig. 2.16b). They can show an organization similar to joints because they often occur in parallel groups or sets. They are not joints, however. Deformation bands commonly occur as mutually cross-cutting sets of conjugate shear fractures (Fig. 2.16b), and they form only in porous, granular strata with low clay content. They can reduce permeability by one to three orders of magnitude relative to the host rock (Antonellini and Aydin 1994).

2.5 Other Fracture-Related Features

2.5.1 Stylolite. Not every induced discontinuity is a fracture. Stylolites are common discontinuities caused by pressure solution of rock. A stylolite is a surface marked by the accumulation of insoluble residual minerals across which the host rock has dissolved in response to pressure solution (**Fig. 2.17**). Stylolites occur most commonly in carbonate rocks. The accumulated insoluble residue reduces permeability; hence, stylolites act as barriers to flow.

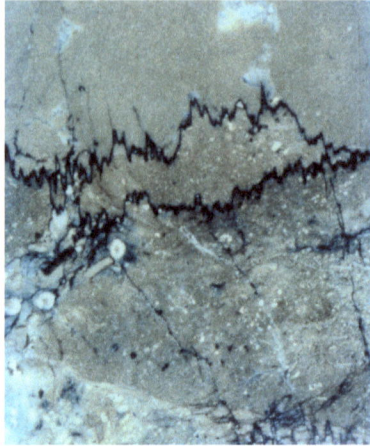

Fig. 2.17—Stylolites in Cretaceous limestone from the Lake Maracaibo region of Venezuela. The height of the "teeth" is a minimum constraint on the amount of rock dissolved as the stylolite developed. Field of view is 8 cm wide.

2.5.2 Breccia. Breccia is a rock composed of coarse, angular, broken rock fragments held together by cement or fine-grained matrix. Breccia is a textural term that describes a rock; it is not a type of fracture. Breccia forms adjacent to some faults **(Fig. 2.18a)**, in caves due to collapse of roof rock (Fig. 2.18b), and in high-energy depositional settings near steep slopes.

2.6 Relationship Among Fractures and Other Geologic Features

How well can we predict orientation(s), fracture density distribution, and other characteristics of fractures when we have little or no data from a reservoir? Is there a useful general model for NFRs? The disappointing answer is: Aside from the guidelines about general fracture characteristics previously discussed, and some reliable lithology-based expectations, there are no generally applicable confident predictors of fracture character in the subsurface. This is a controversial opinion about a controversial topic.

Quantitative statistical approaches to interpolate fracture density throughout a reservoir from borehole-based fracture data are discussed in Chapter 3. The key to predicting reservoir fractures is to seek and understand relationships between observable (e.g., bedding curvature) or predictable (e.g., lithology) features of the reservoir and natural fractures. Although no general model exists, the relationships discussed below are worth considering. However, fracture prediction should be approached warily. The ideas discussed here have a simple and compelling logic, but Earth has a beguiling obstinacy when it comes to honoring human logic.

2.6.1 Fault-Induced Fractures. A common expectation is that fracture density—either extension fractures or secondary shear fractures—is increased near faults. A reasonable body of evidence supports this expectation; however, the practical implication of this effect is controversial. Fault-related fracture enhancement, in the sense of permeability-inducing fractures, has significant potential only in the immediate vicinity of the fault—usually within no more than a few meters (Fig. 2.18a).

Fig. 2.18—(a) Breccia bordering a thrust fault; core diameter is 10 cm. (b) Collapse breccia in carbonate rock, Death Valley National Park (photo courtesy G. Schoenborn).

A nonplanar fault may have a greater tendency to create nearby fractures as rock on opposite sides slide past irregularities. Prediction of subsidiary fracture distributions (fractures induced by the faulting) and hydraulic conductivity related to faults should be based on empirical evidence from other similar faults with the same characteristics in the same reservoir formation.

Predicting the orientation of fault-related fractures is also highly uncertain. Subsidiary fractures in contact with a fault characteristically show an acute dihedral angle with the fault surface.

2.6.2 Fractures and Folds. It is commonly anticipated that when brittle rock is folded, the resulting strain will create fractures, and that fracture density and/or fracture porosity will develop in proportion to the amount of bending (Lisle 1994). Bending can be quantified by curvature C, which in any segment of a 2D cross section of a fold is the reciprocal of the radius of curvature r, so $C = 1/r$ (**Fig. 2.19a**). The 3D curvature is usually defined by Gaussian curvature, which is the product of the maximum 2D curvature and the minimum 2D curvature (Fig. 2.19b):

$$C_{3D} = 1/r_{max} * 1/r_{min}. \quad \dots\dots\dots\dots\dots\dots\dots\dots\dots\dots\dots\dots\dots\dots\dots (2.1)$$

In Fig. 2.19c, the cross-hatched region is the only part of the folded surface area with nonzero Gaussian curvature because both the maximum and minimum curvatures are nonzero.

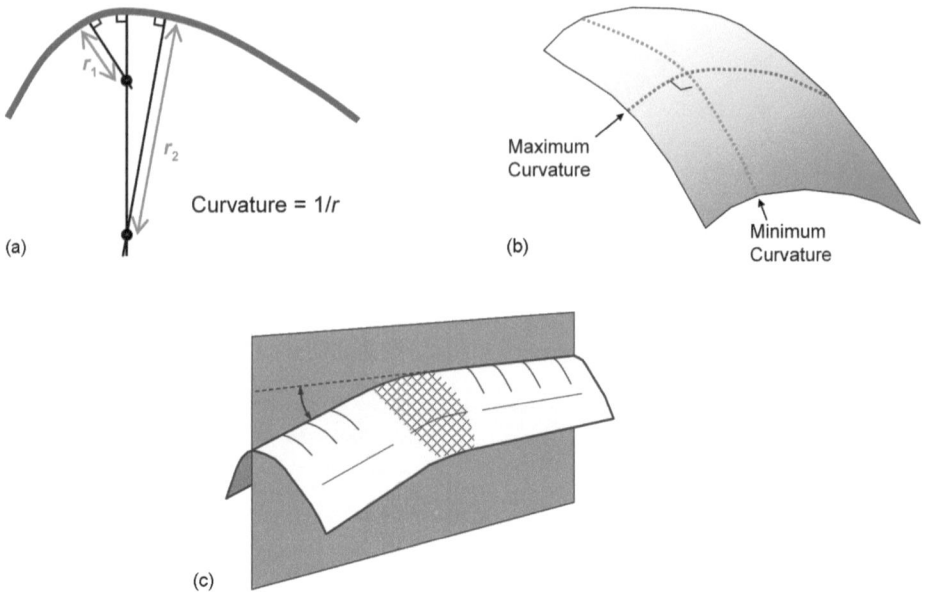

Fig. 2.19—(a) 2D curvature in cross section. (b) Two directions of principal curvature on a 3D domed surface. (c) Hatch pattern shows area of high curvature on a 3D surface.

There has been little validation of the hypothesis that curvature is a predictor of relative fracture density, particularly in subsurface reservoir rock. Several studies have found support for the concept (Lisle 1994; Hennings et al. 2000), whereas others have documented no significant relationship between folding and fracture density (Engelder et al. 1997; Hanks et al. 2004). The curvature hypothesis ignores potentially significant effects such as strain rate, time-dependent rock rheology, folding style, and fracture formation timing. Curvature is no panacea for predicting fractures.

Fracture orientation cannot be confidently predicted relative to folding. This is partially due to the timing of fracture formation; if the fractures formed before or after folding, the stress field orientation may have changed between the formation of the fold and the fractures; therefore, no symmetry should be expected. Even if folding and fracture development are synchronous, the orientation forecast is highly uncertain. The best approach for making such a prediction in a new development is through empirical relationships obtained from local analog studies and regional relationships.

2.6.3 Pore-Fluid Overpressure. Pore-fluid pressure reduces effective stress according to the effective stress law (Jaeger and Cook 1979). Although it is expected that highly overpressured rock will have a greater tendency to develop natural fractures, there has been no empirically based demonstration that this hypothesis is generally applicable.

2.6.4 Lithologic Controls on Fracture Occurrence. Lithology has a definite effect on fracture occurrence. Strong, well-indurated rocks are brittle; hence, they are apt and able to develop extension fractures. Weak or relatively ductile rocks are more likely to deform plastically or to fail in shear, and hence not sustain open fractures **(Fig. 2.20)** (Gross 1995).

Fig. 2.20—Well-indurated dolostone overlying weak mudstone. The strong dolostone contains well-developed joints, whereas the mudstone is nearly unfractured. Miocene Monterey formation, Jalama Beach, California.

Brittleness is not a precise physical property, but is approximately correlative with modulus of elasticity. In an NFR, brittle rocks are more likely to contain fractures than less-brittle lithologies (Gross 1995; Narr 1991).

2.6.5 Stress History. Fractures are a consequence of the total history that a rock has experienced. Taken cumulatively, this history can be viewed in terms of the history of the stress field that has affected the rock plus the evolving physical properties of the rock as it is buried and indurated. The instantaneous stress field is a culmination of rock material properties, thermal stresses (heating during burial, cooling during erosional unloading), remote tectonic stress, weight of overburden, pore-fluid pressure, and local stress effects (e.g., folding or faulting) (Narr and Currie 1982; Engelder 1987). It is little wonder that no simple model for predicting fractures has proved successful.

2.7 Contemporary Stress Field

Stress plays a major role in development of fractures. The total stress field at any location in the Earth's crust is a function of the configuration and movement of tectonic plates, depth, crustal strength, and possibly temperature.

The orientation of the present stress field in the upper crust is widely known and easily determinable from modern image log data (Bell 1990). Stress magnitudes are more difficult to measure because they require specialized tests such as a casing-shoe leakoff test (minifrac), but such data can be reliably obtained (Bell 1990; Zoback et al. 2003). The contemporary in-situ stress field orientation is generally stable and consistent over wide regions **(Fig. 2.21)**.

In most NFRs, the fractures are a relic of an earlier time and stress condition. So the fractures may have developed under very different conditions of stress from those present in

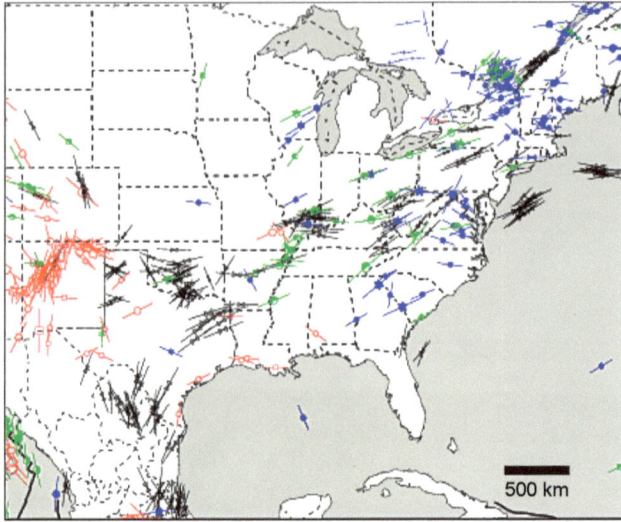

Fig. 2.21—Maximum horizontal in-situ stress direction in the eastern U.S. Note the overall consistency in orientation across broad areas (World Stress Map Project 2004).

the reservoir today; hence, the reservoir fractures may have no genetic relationship with the contemporary stress field. But knowledge of the contemporary stress field can be of practical importance for management of an NFR.

The contemporary stress field controls the orientation of induced hydraulic fractures. Knowing their propagation direction relative to the natural fracture system can lead to improved well-completion planning. Additionally, injected-fluid communication anisotropy in NFRs as well as conventional reservoirs tends to be oriented preferentially in the direction of the maximum horizontal principal stress (Heffer and Lean 1991). This suggests that fractures oriented normal to the contemporary minimum stress direction—those being held shut by the least stress—are statistically the most hydraulically conductive fractures.

2.8 Conclusions

This chapter has provided a geological description of natural fractures, their typical characteristics, and some variations. There are two principal fracture types in the crust: joints and faults. They have distinctly different modes of origin and occurrence. Their characteristics are understood generally, but they are always difficult or impossible to predict. The unpredictable nature of fractures—especially their spatial location and size—poses continued challenges for those who wish to exploit an NFR. The first-order impact of fractures on reservoir permeability, and the great variability of fracture systems in reservoir rock, leads to extreme heterogeneity in reservoir production behavior.

Chapter 3

NFR Characterization

3.1 Introduction

"Fractures are interpreted on the image logs, the simulator is upgraded and ready .... now what do I do?"

This is a common dilemma among those working with NFRs; an unqualified list of fracture occurrences is the basis for running highly idealized simulation software that requires as input grossly generalized, volumetrically distributed permeability information. Bridging the gulf that lies between data acquisition and reservoir simulation, or reservoir management in general, is the focus of this chapter on NFR characterization.

Issues that should be addressed through the course of an NFR characterization often include

- Recognizing which of the many fractures in a well are important.
- Knowing how to integrate diverse data sources (PLT in one well, image log in another, core from a third).
- Deciding what strategy to use for interpolating fracture information through a volume.
- Knowing how to translate and simplify fracture information for a simulator.

NFR characterization is an evolving technology. Today's best responses to these issues will be obsolete tomorrow. Different workers have different approaches; none are ideal, and none are universal. There are never enough data, primarily because of the difficulty of obtaining a representative sample in the extremely heterogeneous medium presented by fractured rock. The one aspect of NFR characterization that is not likely to change is the need to integrate skills from both the geologist and the reservoir engineer to address the diverse and uneven mix of geologic and dynamic data involved. This chapter presents strategies and ideas for addressing these multifarious problems, beginning with geologic description and proceeding to incorporation of dynamic data, arriving finally at integration of all to maximize our understanding for effective reservoir management.

3.2 Detecting Fractures in the Well

Geologic data are those elements of reservoir description that address the geometry and architecture of the fracture system—essentially, the plumbing that holds or transports the fluids. Our understanding of the fracture system is acquired from direct observation of core, from direct wireline-based sensing of the fractures, from the effects of fluid movement within the fracture system, and from knowledge gained from outcrop and producing field analogs. This learning process requires first that we establish that fractures are important, usually on the basis of well performance (see Chapter 1), then use an integrated mix of data to qualify which fractures are important. Modern NFR characterization generally builds around fractures seen in core and image logs.

3.2.1 Core. Core and image logs are the most direct sources of geological detail about fractures in the reservoir. Core provides the most detailed information **(Fig. 3.1)**. From core, we can see unambiguously the relationship between specific fractures and the reservoir rock. Core can be used to determine the origin, geometry, and occurrence of fractures in the reservoir, as well as provide material for routine and advanced engineering analyses.

Fig. 3.1—Tall vertical fracture in a 10-cm-diameter vertical core. The core is continuous from upper left to lower right. The fracture (arrows as guide) changes character downward from a single plane to two parallel fractures to multiple fractures and incomplete core recovery, then back to a single fracture that terminates downward at a bedding-parallel stylolite. Mississippian dolostone, Wyoming.

Core also provides a unique source of information about geochemical modification to fractures, such as the timing of fracture development relative to diagenesis of the reservoir. This information can be critical to understanding when the fractures formed, where they are most likely to occur, and the degree to which they may be sealed by secondary mineralization or enhanced by the action of corrosive fluids (**Fig. 3.2**).

Fracture orientation provides information about fracture-induced flow anisotropy. Fracture dip relative to the core axis and to bedding, as well as relative orientation of fractures, should be measured. Core can be oriented with respect to azimuth by using a downhole camera-compass assembly during coring (Nelson *et al.* 1987). Core can be oriented using standard structural geology methods if the well is not perpendicular to the core and if both the borehole orientation and bedding orientation are known, as from a dipmeter log or structure-contour map (Ragan 1985). A third method for orienting core is by having a specialized laboratory measure the paleomagnetic field in the rock, from which geographic orientation can be determined (Lackie and Schmidt 1993; Hailwood and Ding 1995). A fourth method for orienting core is to make a direct comparison between the core and corresponding features on an image log that records accurate orientation information (Thompson 2000).

Fracture aperture and height should be measured in core. Both are needed to compute fracture density, fracture porosity, and other characteristics. Aperture can be quite variable, and although its measurement is often suspect, a poor measure is better than none. It is important to distinguish open aperture (present-day pore space) from kinematic aperture (how far the fracture walls have moved apart). For example, the fracture in Fig. 3.2b has a kinematic aperture of 1 cm, but its open aperture is zero.

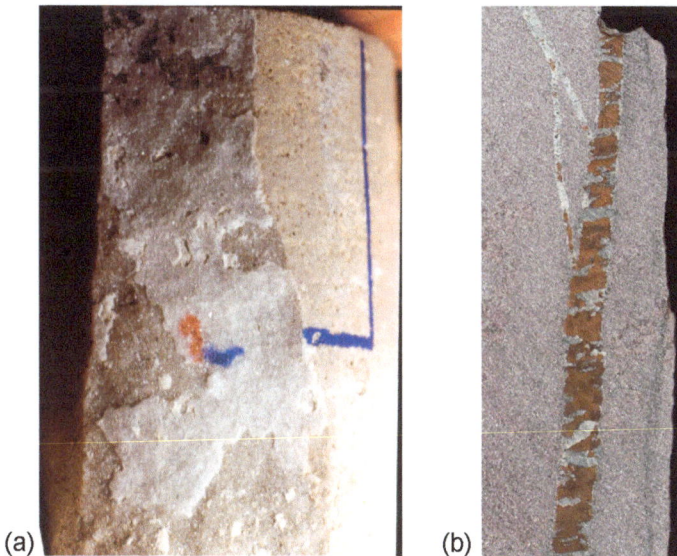

Fig. 3.2—(a) Two episodes of mineralization on an extension fracture: white calcite obscures/covers crystalline quartz (10-cm diameter core). (b) Cross section of 1-cm-wide fracture that was bridged by light quartz crystals then filled completely by dark, iron-rich calcite, Cretaceous sandstone, Sierra Madre Oriental, Mexico (photo courtesy S. Laubach and R. Reed, U. of Texas Bureau of Economic Geology).

Cores are commonly sliced to reveal a clean, flat surface. If possible, fracture descriptions should be done from the whole core, before it is sliced. If it has been sliced, then the interpreter should examine each of the core pieces. Most cores come from vertical wells, parallel to vertical fractures—the worst possible orientation to intersect (sample) fractures (**Fig. 3.3**). Consequently, all available sample material should be examined to ensure that none of the scant data are overlooked.

Those unfamiliar with core description might fail to understand how to distinguish between natural and induced fractures in core. However, this distinction can be made for most fractures with a high degree of confidence (Kulander et al. 1990). Both induced and natural fractures have telltale signatures that divulge their origin. Natural fractures can show mineralization, weathering, or symptomatic surface markings including slickensides and other surface markings that might suggest a point of origin beyond the region of the well. Induced fractures can show geometric symmetry with respect to core, a twisted or curved shape, surface markings indicating a point of origin at the edge of the core, or propagation along the core axis (**Fig. 3.4**).

3.2.2 Image Logs. Image logs are today the most commonly used "direct" source of information about subsurface fractures. Image logs do not replace core, although there is some overlap in the information about fractures they supply. Image logs and core each provide unique visual information about an NFR.

There are two common types of wireline image logs that collect completely different data to create images of the interior of the borehole. Resistivity-based image logs measure small differences in electrical resistivity by dragging arrays of electrodes along the borehole wall to generate an image. Acoustic imaging logs use active downhole ultrasonic transducers that emit and collect reflections from the borehole wall to map subtle variations in shape of

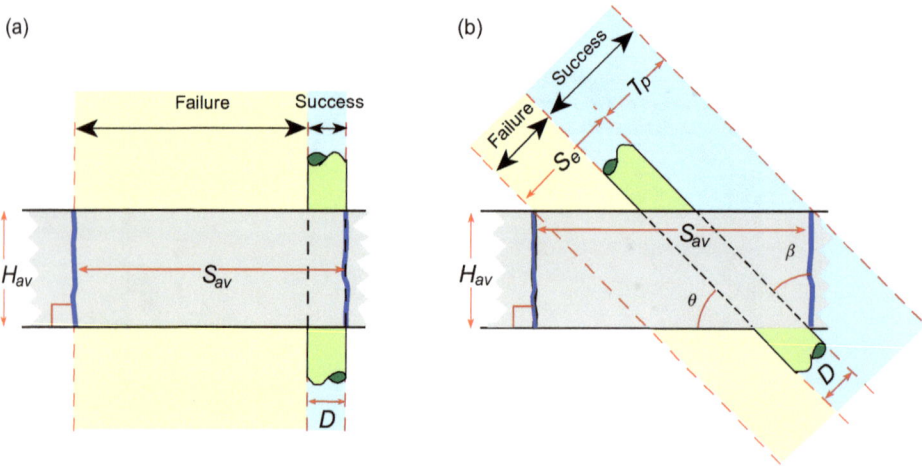

Fig. 3.3—Cross sections showing how borehole-fracture geometric relationships control the probability of fracture intersection across a stratigraphic thickness equal to H_{av}, where "success" vs. "failure" refers to fracture intersection. (a) Borehole is parallel to fractures and normal to bedding. (b) Two-dimensional case of the borehole at an angle to fractures and bedding (after Narr 1996; AAPG © 1996; reprinted by permission of the AAPG, whose permission is required for further use).

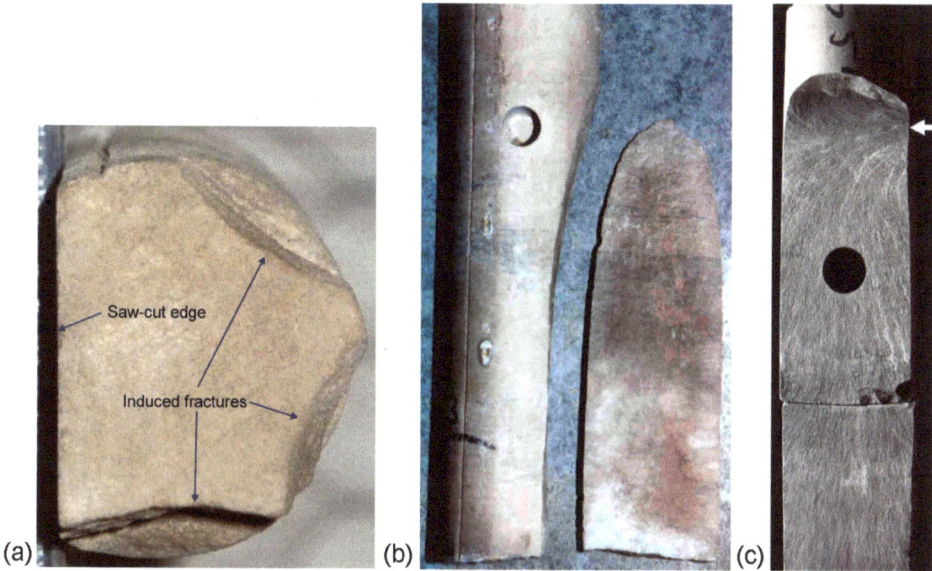

Fig. 3.4—Induced fractures in core (10-cm core diameter). (a) End-on view of core showing fractures curved and symmetrical relative to core. (b) Petal-centerline fracture showing characteristics: concave-down, curving to parallel core axis. On left is cross-section profile; right is view of fracture surface. (c) Plumose surface structure shows fracture initiated at edge of core (arrow), then propagated downward along core. Compare to natural fractures in Figs. 3.1, 3.2, and 3.5.

the wellbore, from which an image is created. Each type of image log has advantages and disadvantages, as summarized in **Table 3.1**. Most workers find resistivity imaging logs superior to acoustic imaging logs, although this is circumstance-dependent. Some logging tools are configured to acquire both ultrasonic and resistivity imaging data, to maximize the advantages of both methods.

Imaging logs are the most useful source of data about location and orientation of reservoir fractures. They are typically interpreted by unwrapping the digital image and displaying it on a flat surface. **Fig. 3.5a** shows a resistivity log in this common, flat representation.

TABLE 3.1—COMPARISON BETWEEN RESISTIVITY AND ACOUSTIC LOGS	
Resistivity Image Log	Ultrasonic Acoustic Image Log
Data collected by 4, 6, or 8 mechanical pads; hence, data are from a spiral striping of the borehole.	Full 360° image around borehole.
Does not function in some oil-based mud.	Functions in oil-based mud.
Generally higher resolution, broader dynamic range, and more sensitive to subtle lithologic variations than acoustic image log.	Sensitivity to lithology varies, based on acoustic properties.
May be used for aperture computation.	Excellent for detailing borehole shape.

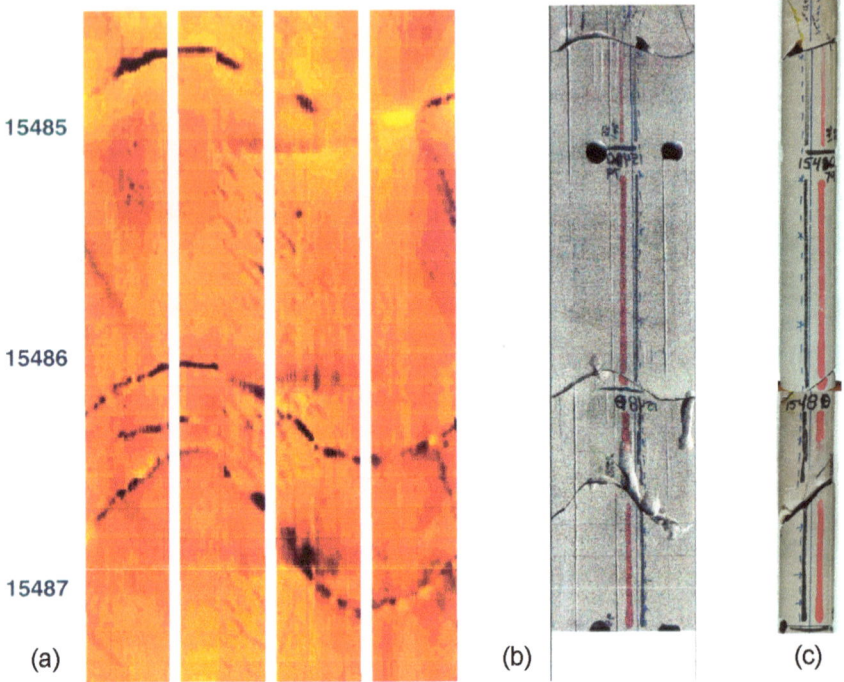

Fig. 3.5—Open fractures in (a) resistivity image log (dark sinusoids), (b) image of core from same interval unwrapped, and (c) image of whole core (courtesy Frank Lim).

The intersection of a circular cylinder (wellbore) with a plane (fracture or bedding surface) forms an ellipse that, when unwrapped and flattened, takes the shape of a sinusoid (Fig. 3.5a). Juxtaposed alongside the image log is a 360° core image of the same interval that was similarly unwrapped. A conventional photo of the core is in Fig. 3.5c.

Although image logs show many features evident in core, some narrow-aperture fractures escape detection. Conversely, whereas core can sometime miss the most significant fractures due to core-recovery losses in highly fractured intervals, image logs will usually capture highly fractured zones without difficulty (Haller and Porturas 1998; Thompson 2000).

In water-based mud, resistivity image logs are able to distinguish open fractures, which are filled by conductive mud filtrate (**Figs. 3.5a and 3.6a**), from sealed fractures that are filled by usually-resistive mineralization (Fig. 3.6b). In oil-based mud, the mud filtrate in open fractures is resistive; if image logging is successful under these conditions, then distinguishing between open and sealed fractures can be challenging. Such circumstances generally require the use of other fracture indicators to differentiate open from sealed fractures (see below). Ultrasonic imaging logs can detect open fractures (Fig. 3.6c), but healed fractures often do not affect the shape of the borehole, and hence do not show.

Image logs are excellent for determining fracture orientation. Logs are generally examined with software that allows the interpreter to quickly fit a sine curve to fractures, bedding, etc. Wellbore survey information and tool orientation are used to compute the orientation of the interpreted planes.

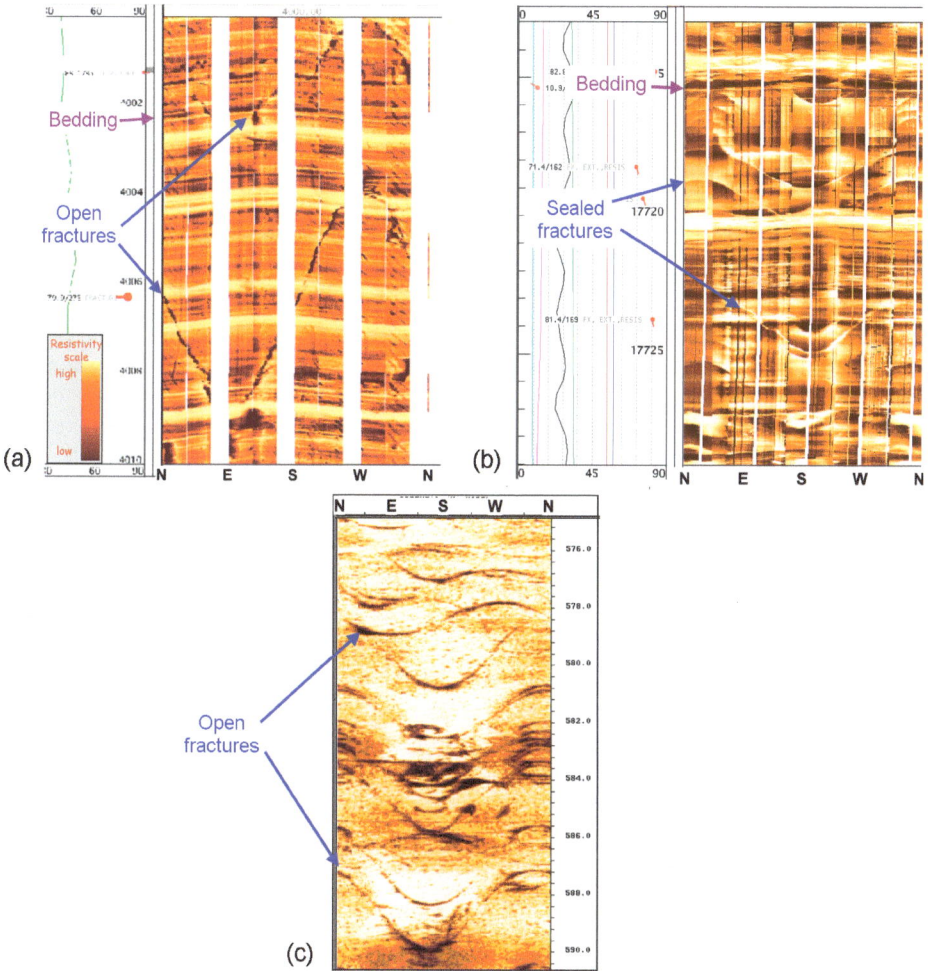

Fig. 3.6—(a) Open, conductive fractures on resistivity-based image log in water-based mud. (b) Sealed, resistive fractures on resistivity-based image log in water-based mud. (c) Open fractures on ultrasonic acoustic log (courtesy Steve Hansen).

Fracture aperture can be computed from resistivity image logs (Luthi and Souhaité 1990). The absolute value of computed apertures has been disputed, but the computed apertures do provide a fairly consistent relative ranking of aperture sizes within a given well, although well-to-well comparisons can be uncertain. Aperture values, in addition to fracture spacing, are needed to compute fracture porosity. If core is available from the same well, or Stoneley-wave aperture determinations (see below), they can be used as an independent check on the image-log apertures.

Apparent fracture height can be measured from image logs. This measure generally underestimates the height of tall fractures that tend to cross the borehole, but the measured height can place a constraint on fracture dimensions, and measured height is used to compute fracture porosity and fracture density. In strongly deviated wells, if the wellbore is nearly parallel to the fractures, intersected-fracture length can be measured. Regrettably, no downhole tool typically provides useful measures of the true height or length of large fractures—those that are of greatest interest. Their dimensions may require estimates based on observations of fracture character at outcrop analogs.

Borehole image logs can supply information on the direction of the present-day in-situ stress field. The stress field can often be determined, whether in an NFR or a conventional reservoir. In vertical wells, the stress field direction can be determined from drilling-induced tensile fractures, which are small hydraulic fractures—often only a few centimeters high—in the borehole wall. If the borehole parallels the plane containing the minimum principal in-situ stress, as is common for vertical wells, the induced fractures strike parallel to the maximum stress direction. Borehole breakouts are distinctly different features that also can be used to interpret the stress direction. Breakouts are oriented zones of enlargement of the borehole caused by compressive-stress-induced failure of the wellbore. The maximum stress direction within the plane normal to the borehole is perpendicular to the direction of breakout. In deviated wells, the stress direction interpretation can be complex (Deeg 1998; Zoback et al. 2003).

3.3 Indirect Fracture Indicators and Validation

Core and image logs provide the most precise and detailed data about fracture occurrence and geometry, but often the most important data for reservoir management comes from other sources. If unqualified, the fracture data from core and image logs can present a confusing collection of fracture statistics that may provide little help with our understanding of dynamic reservoir behavior. When we combine the observed fractures with dynamic information from other sources, our knowledge of the dynamically coupled fracture system is significantly increased.

Various tools have proven useful as indicators of fractures. When we treat them in conjunction with the core or image-log fracture information, they are particularly revealing.

Two questions can be posed regarding fractures seen on image logs (Beur and Trice 2004):

1. Is the fracture open?
2. Is the fracture connected to an extensive, hydraulically conductive network?

The first question is relatively easy to answer; the answer to the second generally bears a high degree of uncertainty. The fractures of particular interest are the "effective fractures," which are able to transport fluid in sufficient quantity to significantly affect flow to a well, or within a reservoir. Effective fractures must answer satisfactorily both of the questions posed here. The following tools help form a basis for identifying effective fractures from well data.

3.3.1 Stoneley Wave Log. Full-wave sonic data from a wireline tool can be processed to obtain the Stoneley (or tube) wave reflectivity response, which has an amplitude attenuation that relates to the aperture of open fractures (Hornby et al. 1989; Luthi and Souhaité 1990). The tool is subject to false positive readings from vugs and washout zones, so using it in conjunction with an image log or core is preferred (**Fig. 3.7**).

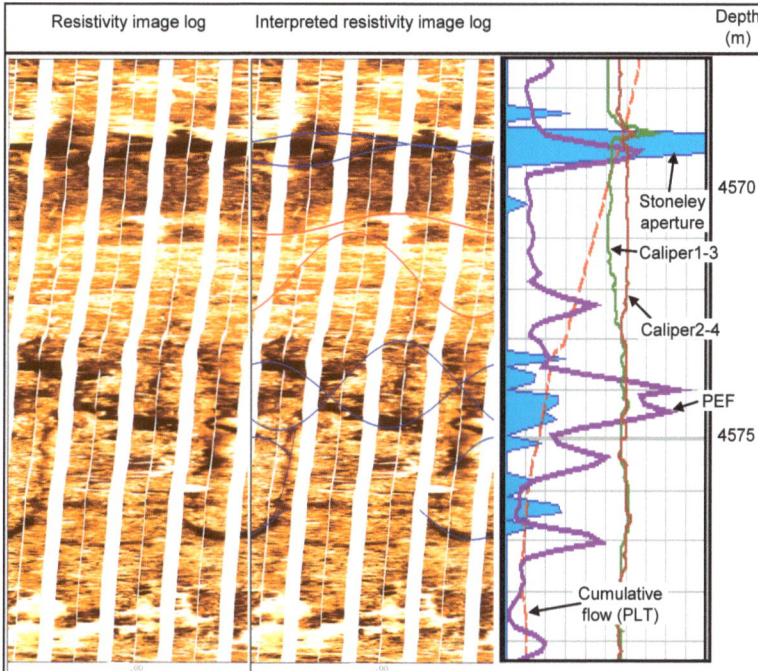

Fig. 3.7—Resistivity image log showing a complex array of fractures. Other wireline logs (right panel) indicate that several fractures are open and effective. The cumulative flow on the PLT production log builds gradually across this interval. Generally, the PLT log does not indicate the precise depth of fluid influx, but has an uncertainty of several meters. There is a slight misalignment in depth among the logs; for example, the Stoneley spike at 4569 m is slightly below the caliper spikes and above the peak PEF spike. Example is from a limestone reservoir, well T-463, Tengiz oil field, Republic of Kazakhstan (courtesy Tengizchevroil).

3.3.2 PEF Log. The photoelectric effect (PEF or PE) log measures the photoelectric response of backscatter electrons, which is used to correct with density log readings. It can be used to recognize open fractures in the presence of barite mud. If a fracture is open and filled with barite mud, the PEF log may spike, clearly showing that the fracture is open. The size extent of the fracture cannot be determined. The PEF response is particularly useful in the presence of oil-based mud, in which open fractures cannot be identified from a resistivity image log alone. PEF spikes occur in the presence of vugs and borehole rugosity, so the PEF log is best used only as confirmation of fractures identified from other methods (Fig. 3.7).

3.3.3 Porosity Logs (Sonic or Density). Reservoir-wide fracture porosity is usually very low (< 0.1%) but can spike higher where fractures are adjacent to the borehole wall, especially if fractures have large apertures. Porosity logs are best used as adjuncts to other tools, especially image logs, to recognize open fractures. They should not generally be relied upon for fracture porosity determination.

3.3.4 Production Logging Tool (PLT). Production logs that are run while the well is flowing allow the interpreter to positively identify which fractures are contributing flow into the wellbore, as well as the amount of fluid that each fracture is conveying. The PLT is the best tool for determining whether a fracture has the capacity to flow significant fluid into the wellbore and by inference, whether it extends a substantial distance into the reservoir. PLT logs generally have two functions: as a flowmeter and as a temperature monitor.

The flowmeter shows an increase of flow into the wellbore as it is pulled past a flowing fracture (Fig. 3.7). The flowmeter is often relatively insensitive to the exact entry point of flow into the wellbore.

The temperature log shows where fluids are entering the borehole on the basis of their heating or cooling effect. This is particularly evident in gas reservoirs where the cooling associated with gas expansion in the lower-pressure wellbore depresses the temperature local to the entry point. A plot of the temperature derivative with respect to depth enhances the visibility of an influx point **(Fig. 3.8)**.

3.3.5 Lost Circulation. Drilling mud can move into fractures encountered during drilling, sometimes quickly draining most of the fluid in the annulus, with disastrous results. More commonly, the amount of mud lost is moderate and can be controlled by the addition of lost-circulation material to the mud. This material clogs fractures and thereby eliminates

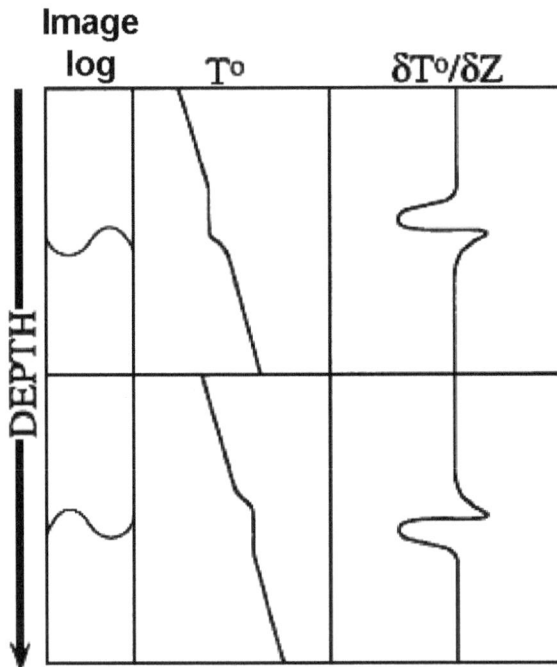

Fig. 3.8—Schematic temperature response from effective fractures, from Barton et al. (1995). The upper example illustrates entry of relatively cold fluid, and the lower example shows entry of relatively warm fluid. The curves on the right are the differential of temperature with respect to depth.

losses. NFRs can present a delicate balance for drillers. However, data on lost circulation can help in the recognition of effective fractures.

Drillers report lost circulation and the rate of loss, usually in gallons per hour. Loss into permeable matrix generally builds gradually, whereas the loss to a fracture will start abruptly **(Fig. 3.9)**. Driller's logs are typically not precise, but are useful nonetheless. The depth of stepwise increases in lost circulation can be compared to fractures recognized in other data sources. Highly accurate lost-circulation monitoring may provide a method for resolving fracture size and effective fracture apertures based on the rate of mud lost (Dyke et al. 1992; Verga et al. 2000).

Use of lost-circulation data has several potential pitfalls. After lost-circulation material is added to the mud system, newly encountered effective fractures may clog quickly and not be recognized. Alternatively, the dynamic hydraulic changes that occur in a well during drilling can cause a previously blocked fracture high above the level of the bit to unclog, resulting in a lost-circulation event. This can be mistakenly identified as a new fracture at bit depth. Using lost-circulation data in tandem with core or image log data will help to mitigate these potential misinterpretations.

A final note about lost circulation: Drilling mud can clog fractures and adversely affect the wettability of fracture surfaces, causing severe damage to the reservoir. Reversing this damage may not be possible, so despite its usefulness as an indicator of effective fractures, losing circulation is not to be encouraged. Indeed, underbalanced drilling is favored in many NFRs to avoid losing drilling fluid into fractures.

3.3.6 Gas Shows in Mudlog. Abrupt increases in gas shows, or spikes, or kicks in the mud while drilling, especially if they correspond with a known fracture or an interval of low matrix quality, may indicate an effective fracture.

3.3.7 Mechanical Indications of Fractures. Several mechanical effects can indicate the presence of open fractures, although size is not indicated. "Caliper" logs can show short (height < 0.5 m), abrupt spikes caused when a fracture wall breaks off during drilling. These are not the stress-induced borehole breakouts that tend to occur over long intervals of borehole (Bell 1990), but rather are confined to the small area where a natural fracture and bore-

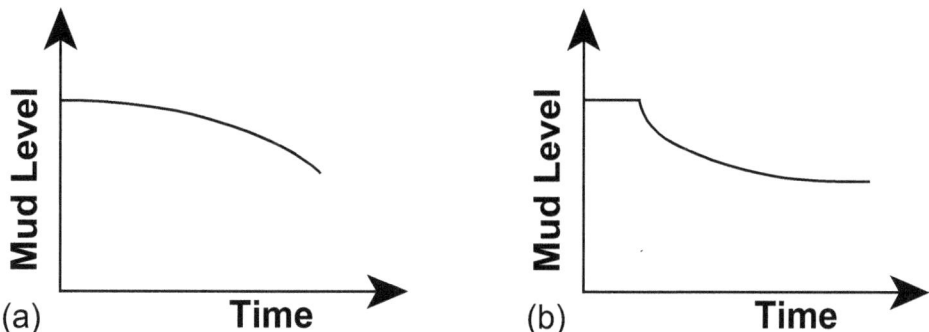

Fig. 3.9—Mud loss during drilling monitored by level of mud pit. Mud loss can show a continuous increase (a) caused by high matrix permeability, or a sudden increase (b) indicative of an effective fracture (from Verga et al. 2000, after Dyke et al. 1992).

hole intersect (Fig. 3.7). Orientation of the caliper spike should not be trusted because the borehole wall can fail in various directions depending on the fracture-borehole geometry. The fracture orientation should be obtained from the image log or core.

When a drill bit encounters an open fracture, the response may be evident as an increased rate of penetration or a decrease of torque on the bit. Rarely, but spectacularly, the bit drops into a cavity in the rock. This is most common in carbonate rocks, where solution enlargement of fractures by corrosive fluids is common.

3.4 Orientation and Spatial Organization of Reservoir Fractures

After subdividing fractures according to their effectiveness, or other characteristics of interest, the organization of the fractures can be investigated. This consists of understanding the orientation of fractures, classifying them into sets, and determining their density variation spatially and with respect to lithology or other geologic controls. Data should be displayed on a variety of analytical plots for the purpose of looking at it from different perspectives, to gain insights, and to understand relationships that might not be evident from any single point of view. This is data exploration.

3.4.1 Orientation. Several conventional diagrams are useful for defining the orientation distribution of fracture populations. A rose diagram is a radially symmetric circular histogram of frequency as a function of fracture strike direction. For example, the rose diagram in **Fig. 3.10a** shows a predominant fracture strike that is nearly N/S. A rose diagram shows only a 2D distribution; it does not show fracture dip.

A stereonet is a diagram for representing the full 3D orientation of structural features such as fractures (Figs. 3.10b and 3.10c). For the construction of a stereonet, a fracture is circumscribed according to its orientation inside a lower hemisphere, and the line perpendicular to the fracture (its pole) is projected from the top center of the fracture plane down onto the hemisphere. This point is then projected upward onto the horizontal plane that forms the top of the hemisphere. The stereonet in Fig. 3.10b shows that the fractures dip ~70° west, based on the cluster of points near the east edge of the circle; this dip information cannot be determined from the corresponding rose diagram. Points on the stereonet are individual fracture orientations; they have been contoured to accentuate orientation clusters. Additionally, other orientation data such as bedding can be added to the stereonet easily. Fig. 3.10d is a block diagram illustrating these fracture data.

Fractures should be assigned into sets, which are treated separately when computing fracture density or other parameters. Fracture sets are usually defined based on orientation, which is judged from inspection of orientation diagrams.

3.4.2 Spatial Organization. Orientation diagrams are for examining orientation of planes (e.g., fractures and bedding), but they are not designed for analysis of the spatial position of these planes. For this, a number of other graphical tools are useful. The tadpole plot, which is commonly used to display bedding orientation on dipmeter logs (logs that measure bedding dip in wells), can also be used to display fracture data as a function of depth in a well **(Fig. 3.11a)**. On these plots, the "head" of the tadpole shows the depth (y-axis) and dip magnitude (x-axis) of the fracture; the "tail" shows the fracture's dip direction (up is north). In most 3D modeling software, fractures can be displayed as discrete planes in space along the trajectory of a well (Fig. 3.11b). Fracture attributes such as aperture, min-

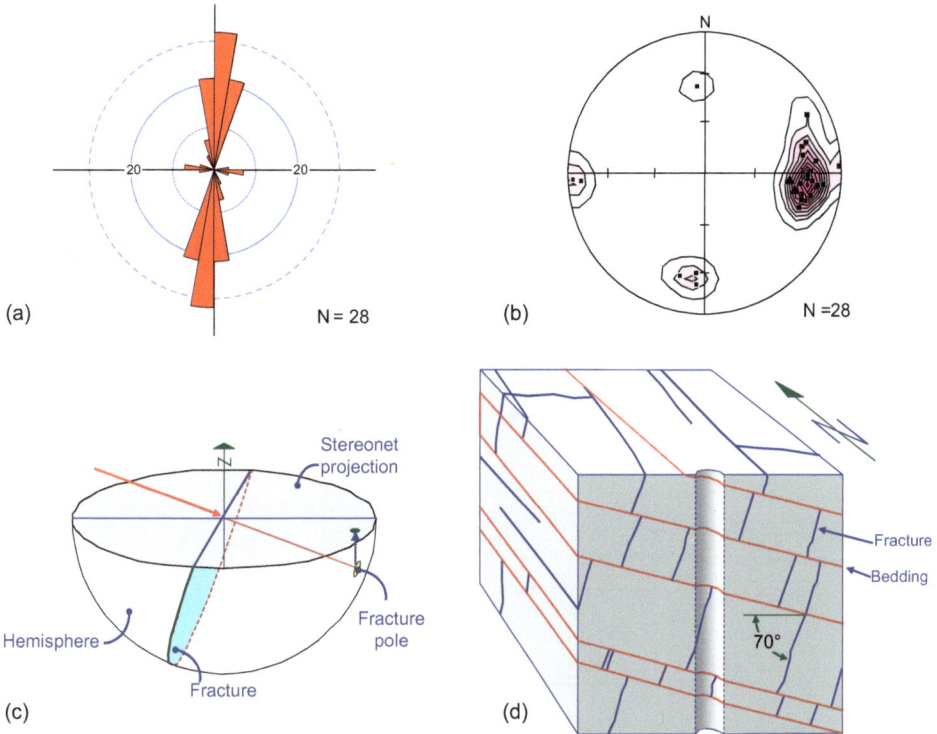

Fig. 3.10—(a) Rose diagram and (b) contoured stereonet showing orientations of fractures encountered in a well. (c) Schematic illustration of elements of stereonet construction for a single fracture dipping steeply toward WNW. (d) Block diagram illustrating schematically the data shown in (a), (b), and (c).

eralization, and effectiveness can be displayed by different colors, for example, which can lead to a better spatial understanding of fractures within the reservoir volume.

3.4.3 Fracture Density and Fracture Porosity. Most of the characterization described thus far has focused on understanding the behavior of individual fractures or orientation-defined groups. A common method for homogenizing the fracture information is to compute spatially localized fracture density for different fracture sets. Even if the ultimate goal is to build a discrete-fracture model, the basis for quantifying the spatial variation in degree of fracture development is commonly with a fracture-density distribution. Fracture porosity can also be used as a basis for quantifying degree of fracturing.

Fracture density may provide a quantitative basis for interpolating fractures throughout a volume. Although fracture density has a realistic physical basis—it is the fracture surface area per unit volume of rock—in the unseen volume of the reservoir, it is a statistical generalization: an estimate of a parameter that has high variability and high uncertainty.

Fracture surface area is easily computed from fracture data collected from core or image logs. The volume of the sampled interval V_r is assumed to be a right-circular cylinder (this is the volume of the core or, if based on image log data, the borehole), the area of an indi-

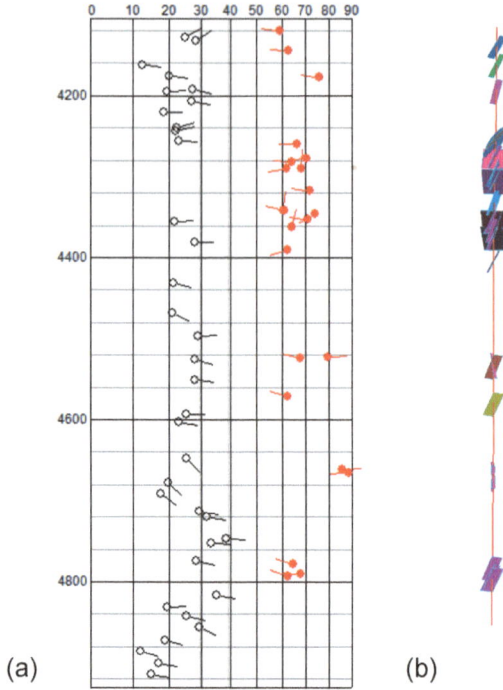

(a)   (b)

Fig. 3.11—(a) Tadpole log showing fracture (solid) and bedding (open) orientation with measured depth in a well. (b) Fracture planes displayed along a well trajectory using 3D visualization software.

vidual fracture A_f can be treated as an ellipse within that volume (upper fracture in **Fig. 3.12**), and it is approximately correct for a truncated fracture (lower fracture in Fig. 3.12).

$$V_r = \pi \left(\frac{D}{2} \right)^2 h_r, \qquad\qquad\qquad\qquad\qquad\qquad\qquad\qquad\qquad\qquad (3.1)$$

where D is the diameter of the core or, in the case of an image log, the borehole.
   The fracture density, d_f, is:

$$d_f = \frac{\sum_{i=1}^{n_f} A_{fi}}{V_r} = \frac{\sum_{i=1}^{n_f} h_{fi}}{D h_r}. \qquad\qquad\qquad\qquad\qquad\qquad\qquad\qquad (3.2)$$

   The units of fracture density are L^2/L^3. This can be reduced to $1/L$, and, thus, for a set of parallel fractures, L is equal to their average spacing (perpendicular distance between fractures). Hence, the computation of fracture density provides an estimate of average fracture spacing even for cases in which the borehole is parallel to fractures, as in the common circumstance when both fractures and borehole are vertical (Narr 1996).

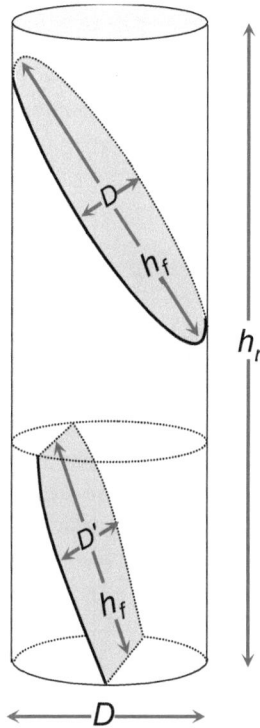

Fig. 3.12—Borehole (or core) of diameter D and height h_r crossing fractures with a visible height of h_f. The edges of the upper fracture intersects the borehole in all directions. Both the upper and lower edges of the lower fracture terminate within the borehole; therefore, this fracture does not exactly match the geometry of an ellipse.

The average spacing obtained from the fracture-density computation can be used for well planning or as an input parameter in reservoir simulation. For these and most other applications, the effective fractures should be used. They tend to be a small subset of the total fracture population in a well.

Fracture porosity (ϕ_f) is computed by adding fracture aperture a_f to the fracture-density computation:

$$\phi_f = \frac{\sum_{i=1}^{n_f} a_{fi} A_{fi}}{V_r} = \frac{\sum_{i=1}^{n_f} a_{fi} h_{fi}}{D h_r} . \dots\dots\dots\dots\dots\dots\dots\dots\dots\dots\dots\dots\dots\dots (3.3)$$

In most NFRs, the fieldwide fracture porosity is negligibly small ($< 0.1\%$).

The well interval thickness used for the fracture-density or fracture-porosity computation should be thick enough to contain a representative sample of fracture occurrence. Fracture sampling is a function of fracture spacing, borehole deviation, and borehole diameter (Narr 1996). The absence of fractures in an interval does not necessarily mean the interval is unfractured; it simply means that no fractures were sampled. Conversely, the presence of a fracture could be a statistically rare event; the well may intersect several, although their spacing in this zone may be very wide. Increasing the thickness of the interval of evaluation, h_v, is the only way to counter this sampling issue. A representative interval thickness can be tens of meters or more in a thick reservoir. In most cases the well-based sample is below the representative elemental volume (REV) of the fracture system (Gilman 2003). Herein lies the root of the heterogeneity that typifies the behavior of most NFRs.

3.5 Fracture Density Distribution

Understanding and predicting the reservoirwide distribution of fracture density are difficult and important steps toward NFR characterization. Fracture density or fracture porosity, computed for intervals along wells, can be treated like other well log curves (**Fig. 3.13**). Notice that fracture density is not proportional to fracture frequency; the high magnitude of fracture density near the bottom of the interval (*L*) is based on only two fractures, whereas the greater number of fractures toward the top of the well (*U*) has a lower fracture density. This is because the fractures in the lower part of the well are tall and nearly parallel to the wellbore; hence, they have a high fracture surface area. In the upper interval, the fractures crosscut the wellbore, so their cumulative surface area is not great. The probability of intersecting wellbore-parallel fractures is lower than the probability of hitting fractures that crosscut the wellbore. The simple fracture-density definition (Eq. 2) normalizes these competing sampling-bias issues if a statistically representative interval is sampled. Unfortunately, well-based fracture-density calculations are subject to the whims of small-sample-number statistics.

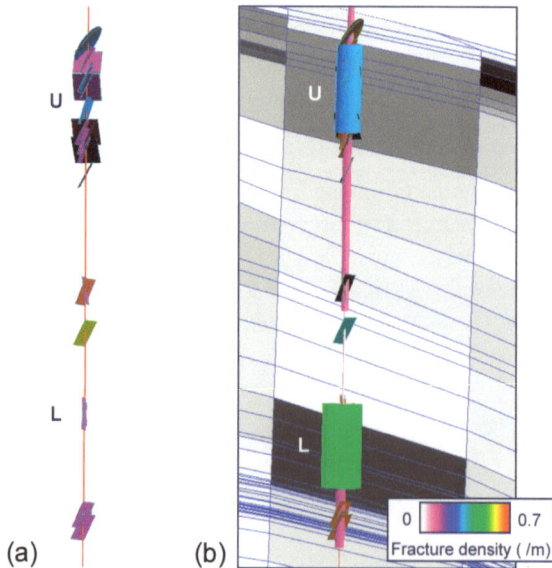

Fig. 3.13—(a) Discrete fractures displayed along a well in a 3D model. (b) Fracture density computed along the well, displayed as spindles, with spindle diameter proportional to fracture density. The simulation grid is the cross-section shown in the background, with fracture density as shading. The right-dipping lines in the grid represent bedding. Compare to the tadpole plot of Fig. 3.11.

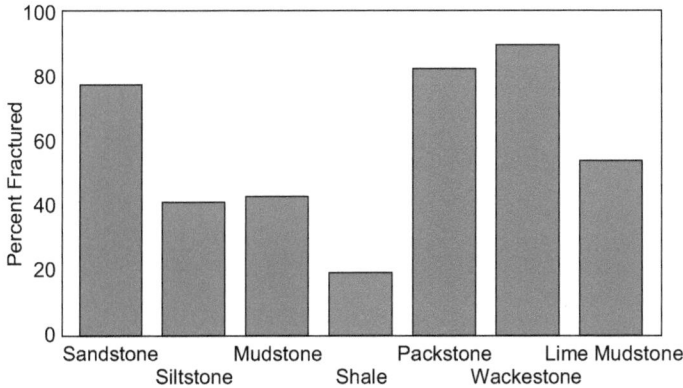

Fig. 3.14—Differences in fracture occurrence as a function of lithology, documented at Bluebell oil field, Uinta basin, Utah (from Montgomery and Morgan 1998; AAPG © 1998; reprined by permission of the AAPG, whose permission is required for further use).

The fracture data should be examined to search for obvious geologic correlations such as relationships of effective fractures and lithology, structural features, or stratigraphic position. For example, **Fig. 3.14** shows, for one field, a relationship between fracture occurrence and lithology.

Some properties, such as matrix porosity, can be interpolated throughout a reservoir volume using established geostatistical methods. Fracture distributions, however, are not known to behave according to conventional geostatistics. Multiple-regression methods, such as neural-net-based routines and discriminant analysis, have been used to predict fracture density throughout a reservoir based on relationships with such variables as matrix porosity, lithology, shale content, fold curvature, and seismic attenuation (Ouenes et al. 1995; Gauthier et al. 2002). This is a two-step process; first, the data are interrogated to look for relationships between fracture density and spatially distributed, potential predictor data at coincidental points along wells ("training" in neural-net jargon). Then, these empirical correlations are used with the spatially distributed predictor data to predict fracture density throughout the reservoir volume.

In their study of a field in North Africa, Gauthier et al. (2002) looked for correlations between fracture density and various other geologic variables (**Fig. 3.15**). They found that some variables are relatively highly correlated with fracture density (dolomite content, for example), whereas other variables, such as seismic coherence, do not correlate. Note that their expression "fracture index," which they define as reciprocal spacing, is equal to fracture density. They used all of the significant drivers, taken together, to predict fracture density throughout a model volume.

A multiple-regression approach produces a prediction based more on objective quantitative statistical correlations than intuition, although the choice of predictive control variables (drivers) can be subjective. A parameter that correlates strongly with fracture density will be more influential in determining its prediction than a poorly correlated variable. This approach implicitly recognizes that fractures result from a complex interplay of geologic factors. A multivariate predictive approach mimics the naturally complex multiple-influence interplay that impels fracture systems in nature.

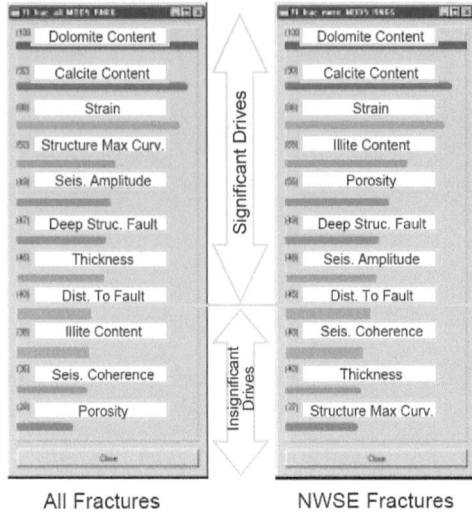

Fig. 3.15—Drivers are variables that correlate with fracture density; hence, they can be used in a predictive model. The variables shown here are those for all fracture sets combined (left) and for just one fracture set evaluated independently (right). Length of bar beneath label shows relative strength of correlation (from Gauthier et al. 2002).

A fieldwide fracture-density model can serve as a basis for an NFR simulation model. If the simulation model is to be a single-porosity effective-permeability model, the magnitude of fracture density can guide the relative magnitude and distribution of permeability multipliers. If a dual-porosity model will be used, the modeled fracture spacing S_f is a function of the reciprocal of fracture density ($S_f = 1/d_f$). The "shape factor" in dual-porosity models, which makes use of fracture spacing, is a mathematical construction; it may be proportional to the geological fracture spacing, but it should not be assumed to equal it.

3.6 Using Analogs in NFR Characterization

Information from analogs is important—in many cases critical—in NFR characterization. Analogs can provide a generalized view that is not readily obtainable during development of an NFR. Also, some data are rarely obtainable from the subsurface, such as bedding-fracture relationships, fracture lengths, and fracture-fracture relationships, so outcrop analogs may provide constraints on relationships and measurements. There are two principal sources of analog information: producing fields and outcrops.

3.6.1 Producing Field Analogs. Analog fields provide constraints on the producing characteristics of geologically similar NFRs. Analog fields can be used as a guide to productivity, to recovery factor, and possibly to characteristics such as production heterogeneity and water influx. When selecting fields to use as analogs, the general process is to select geologically similar fields (same or lithologically similar reservoir, same structural and tectonic setting, same burial depth, same geologic history, same pore-fluid pressure, etc.) from published or other available sources. This is difficult in NFRs because fracture systems are variable, and there is rarely adequate information about the fracture system. The best use of field analogs is to constrain a range of possible behaviors by reviewing the behavior of a variety of analogs. Studying analog fields also leads to new ideas and comprehension of the fracture geology of the subject NFR.

3.6.2 Outcrop Analogs. Outcrops can provide a representative qualitative understanding of the nature of fractures in a reservoir. The outcrop perspective is largely 2D, or a collection of arbitrary 2D views that can be stitched together—either mentally or with the assistance of modeling software—to gain a limited 3D view of a fracture system. The outcrop analog may be the reservoir rock formation, or it may be a rock chosen on the basis of similarity to the reservoir in lithology, stratigraphic setting, rock properties, age, structure, etc. For example, in order to understand variations in fracture density related to the geometry of an oilfield-size fold, Hennings et al. (2000) examined an outcrop of the Frontier formation in Wyoming. In another example, Gale (2002) uses data from outcrop analogs to constrain the distance between major fractures in the Austin Chalk in order to guide the planning of horizontal wells that target such fractures.

Outcrops can provide a clear understanding of how fractures relate to lithology or to stratigraphy. The areal variation of fracture characteristics can be determined if exposures are good. Just as in the subsurface, the important parameters of fracture length and connectivity are among the most difficult to measure. Exceptional outcrops are needed for data on these parameters, but they are worth seeking.

A fracture system exposed at outcrops should be assumed to be different from a fracture system in the same rocks at several thousand meters burial depth. The differences can be profound. As a rock is uplifted to the surface, its stratal overburden is removed by erosion, its temperature cools significantly, tectonic stress is reduced, and pore-fluid pressure changes (Narr and Currie 1982; Engelder 1987). These processes significantly change the stress field on the rock, which can result in extension of existing fractures and development of new fracture sets. Although NFRs commonly contain only a single predominant fracture set, outcrops commonly contain two or more sets of fractures (e.g., Fig. 2.10). Consequently, it is important for observers to attempt to filter the effects of late-acting processes from the natural fractures seen at outcrop.

Establishing the regional context of a subsurface fracture network can provide valuable understanding of a reservoir. Understanding the regional context of an NFR is not really an analog issue, but the collection of data can overlap with outcrop-analog data collection. The focus of a regional study is generally fracture-set orientation and development. If the fracture system observed in an NFR is part of a regional joint set, orientation trends are probably consistent throughout the reservoir. Also, the variability of fracture-set orientation and degree of development in the reservoir may mirror the regional condition. If the fracture-set orientation in the reservoir deviates strongly from regional trends, it may be a local fracture set, and hence less regular in orientation and distribution.

3.7 Using Dynamic Data in NFR Characterization

Dynamic data provide a critical dimension that propels NFR characterization toward the substantial issue of reservoir performance. Dynamic data lead to a better understanding of the behavior of an NFR on a large scale and can be useful for calibrating the geology-based fracture model in preparation for a reservoir simulation model.

Well tests can provide several lines of evidence that bear on the dynamic behavior of the NFR. These include:

- Single-well tests:
  - ○ Well kh (permeability k times measured height of producing interval h).
  - ○ Linear vs. radial flow behavior.
  - ○ Dual porosity.
- Multiwell interference (pulse) tests:
  - ○ In-situ flow anisotropy, k, kh, and ϕh.

As applied to natural fracture systems, conventional well-test analyses (e.g., standard dual-ϕ formulations) are based on ideal mathematical constructions of the reservoir rather than geologically realistic fracture networks. Analyses based on more realistic discrete-fracture-network (DFN) models can improve the calibration between well tests and a geologic model.

3.7.1 Single-Well Tests. *Well kh Ratio.* The ratio of kh_{well} to kh_{matrix}, or FCI, was described in Chapter 1. The primary value of this measure in characterization is to document the magnitude of fracture-related flow and its variability throughout the reservoir. A map of FCI values suggests whether certain portions of the reservoir are more effectively fractured than other portions, and it also provides a semiquantitative indication of the magnitude of permeability enhancement.

The value of kh is complicated in an NFR because flow into a wellbore can be more limited than the height of interval open to production. For example, measured kh may be based on a well test of a 50-m-high completed interval, but within this zone, 95% of the flow is from a 3-m-high interval containing a few identified fractures (e.g., Figs. 3.7 and 3.8). To understand fracture flow, the kh of the 3-m-high fracture zone might be computed separately from the remaining 47-m zone. However, if the entire 50-m interval feeds the fractures, perhaps it should all be considered together. Evaluating an extremely heterogeneous flow system such as this is not routine and should be approached thoughtfully. Consideration of several alternative descriptions of h is usually warranted.

Linear Flow Behavior. The rate of change in pressure during a drawdown or buildup test can give indications of effective fractures. If a well intersects a fracture—either a natural fracture or a fracture induced during well stimulation—flow will be from the fracture to the well, and the fracture will be recharged from the matrix **(Fig. 3.16a)**.

A time period of linear flow may occur when the pressure support is primarily along a fracture connected to the well, which may be recharged by the matrix (Fig. 3.16) or by a connected fracture system. The characteristic drawdown response on a log-log pressure-derivative curve shows a slope of $\frac{1}{2}$ **(Fig. 3.17)**. If fractures connected to a well are of limited extent, the flow response will progress, after a period of time, to radial flow behavior (Fig. 3.16), which shows an exponential shape, evident as a straight line of constant slope on the derivative plot. The simplistic illustrations in Fig. 3.16 are highly idealized finite-length fractures; in actual fracture reservoirs, the complex geometry may affect the well-test pressure response, resulting in derivative curves that are not as simple as these ideal models.

Dual Porosity. In some NFRs, a well test may show both a fracture-dominated response and a matrix-dominated response, which can be considered a dual-porosity system. However, absence of a characteristic dual-porosity well-test response does not indicate that the reservoir is unfractured. A majority of NFRs do not show dual-porosity behavior in well tests.

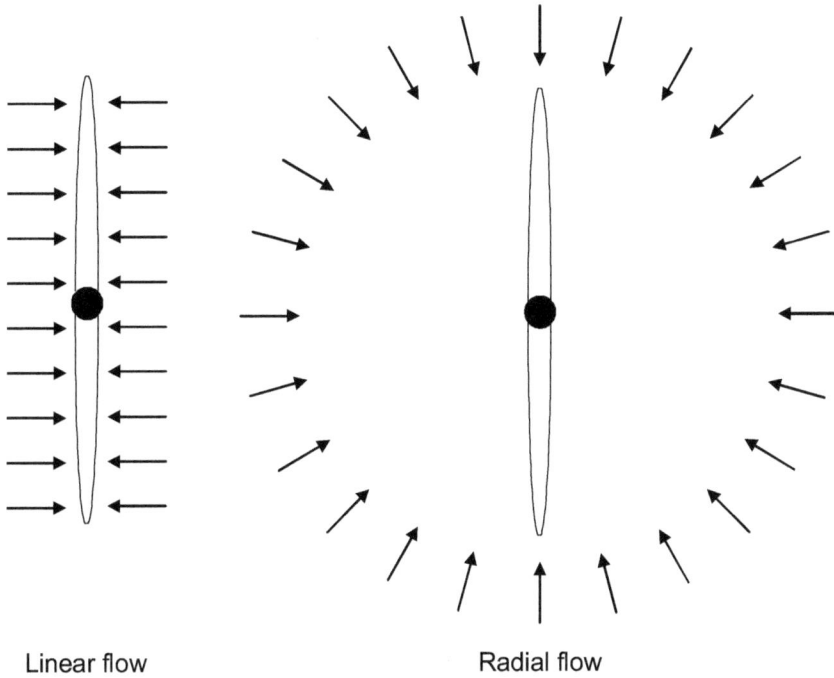

Linear flow   Radial flow

Fig. 3.16—Flow regimes relating to a vertical finite-length fracture in a permeable matrix inter-sected by a well, seen in map view (after Horne 1995).

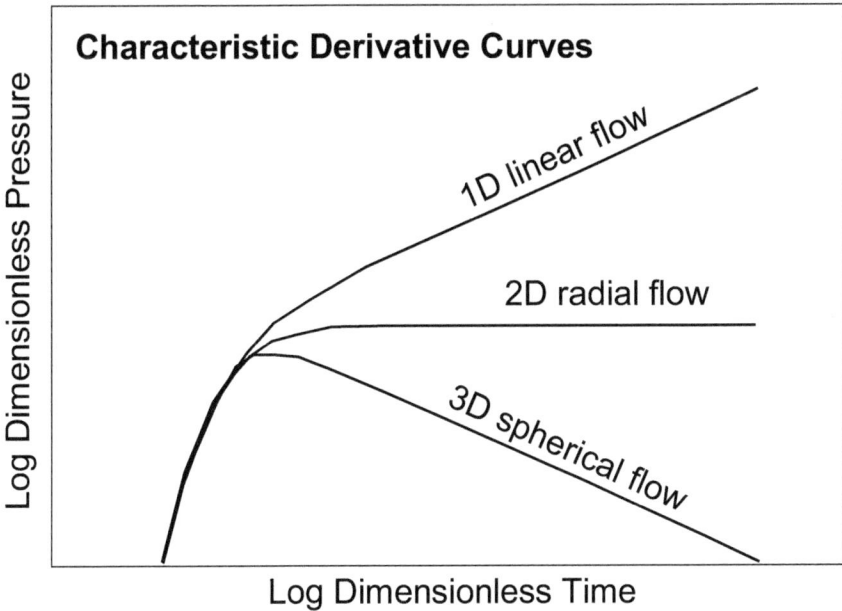

Fig. 3.17—Plot of idealized pressure-drawdown derivatives for a constant-flow test.

A dual-porosity system is a reservoir that presents two distinct pressure responses, one for the fracture network and one for the matrix, as well as a transitional response between these two porosity systems.

Warren and Root (1963) recognized that the drawdown and buildup responses in some NFRs exhibit two parallel straight lines on a plot of log Δp vs. time, the earlier-time line corresponding to the transient flow in the fractures and the later-time line to the transient flow in the total system **(Fig. 3.18)**. The slope of these lines is related to the flow capacity of the formation, and the vertical separation of the two lines is related to the relative storage capacity of the fractures and the matrix.

A typical pressure drawdown response in a dual-porosity reservoir is shown in Fig. 3.18a, where the pressure is plotted against the flow time. The plot is characterized by the first straight line, a transition, and a second straight line having the same slope as the first line. The first straight line represents the fracture system; it is usually of short duration and difficult to detect in practice because of wellbore storage. The second straight line represents the total system behavior. The separation between the two straight lines represents the magnitude of fracture storativity (ω). The location of the transition relative to the time axis relates to the interporosity flow, quantified by the interporosity flow coefficient λ, which describes the flow from matrix to fractures and relates to fracture surface area. Behavior of a buildup test is similar (Fig. 3.18b).

The storativity ratio, ω, is defined as:

$$\omega = \frac{\phi_f C_{tf}}{\phi_f C_{tf} + \phi_m C_{tm}} . \quad\dots\dots\dots\dots\dots\dots\dots\dots\dots\dots\dots\dots\dots\dots\dots\dots\dots\dots (3.4)$$

C_{tm} and C_{tf} represent the compressibility of the matrix and the fractures respectively. A larger difference between matrix porosity and fracture porosity will result in a larger separation between the straight lines of Fig. 3.18 and a resulting smaller value of ω. In some cases the compressibility for matrix and fracture is nearly equal, in which case ω reduces to the ratio of fracture porosity to total system porosity. But the value of fracture compressibility can be much higher than the matrix.

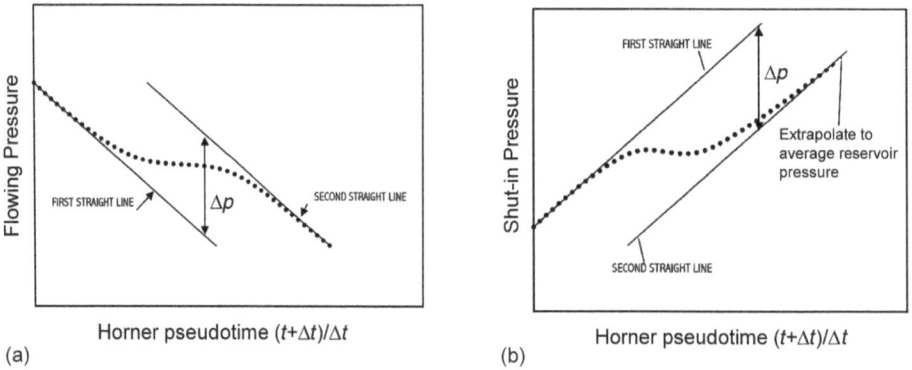

Horner pseudotime $(t+\Delta t)/\Delta t$

(a)   (b)

Fig. 3.18—Pressure response in a naturally fractured reservoir showing ideal dual-ϕ behavior during pseudo-semisteady-state flow. (a) Drawdown test. (b) Buildup test.

Eq. 3.4 can be re-arranged to solve for fracture porosity:

$$\phi_f = \frac{C_{tm}\phi_m}{C_{tf}\left(\dfrac{1}{\omega} - 1\right)} . \quad\dots\dots\dots\dots\dots\dots\dots\dots\dots\dots\dots\dots\dots\dots (3.5)$$

This estimate of fracture porosity from the well test (Eq. 3.5) can be compared with the value obtained from interpretation of individual effective fractures in the well-test interval. The well-test value for fracture porosity represents the effective fractures in bulk, whereas the individual-fracture computation will be divisible into separate fracture sets. However, the dynamic well-test-based fracture porosity will represent a greater volume of reservoir and might be a better basis for a simulation model, even though the value of fracture compressibility is generally not known. Furthermore, a single number is not likely to accurately appraise fracture compressibility for a real fracture system. Rationalizing these independent measures of fracture porosity can help in understanding the uncertainty inherent in estimating fracture porosity.

The second key variable to be derived from a dual-porosity well-test response is the interporosity flow coefficient, λ, which, for a system with matrix-to-fracture pseudosteady-state flow, is:

$$\lambda = \frac{\sigma k_{avg} r_w^2}{k_f} , \quad\dots\dots\dots\dots\dots\dots\dots\dots\dots\dots\dots\dots\dots\dots\dots (3.6)$$

where k_f is fracture permeability and k_{avg} is average effective permeability of the matrix plus fractures, r_w is wellbore radius, and σ is a shape factor that relates to the fracture surface area and its arrangement.

A dual-porosity response on a log pressure vs. log time plot is shown in **Fig. 3.19**, where both dimensionless pressure and its logarithmic derivative are plotted as functions of dimensionless time. The transition of the pressure response from the fractures to the fracture-plus-matrix system will appear as an area of minimum slope in the pressure-time curve, which is more evident in the derivative curve as a trough minimum. When ω is small, the derivative at the minimum is close to zero.

Although dual-porosity well-test analysis presents a clear set of analytical tools, in practice its usefulness is limited. Many NFRs do not show dual-porosity behavior. The response transition from fractures to fractures-plus-matrix often occurs so quickly that it is masked by wellbore storage effects. Furthermore, geometry issues such as the physical connection between the wellbore and a real fracture system, or the organization of the fracture system itself, can markedly impact the well-test pressure response and thereby alter or mask the dual-porosity response.

3.7.2 Multiple-Well Tests.
Multiple-well testing provides a measure of reservoir flow behavior through a large volume of the reservoir. This is typically done using interference (pulse) testing, during which a pressure pulse is created and its response is monitored in off-set wells. Tracer tests, in which a chemical or radioactive tracer is injected and its appearance in offset wells is monitored, are commonly used in groundwater studies but less frequently in petroleum reservoirs.

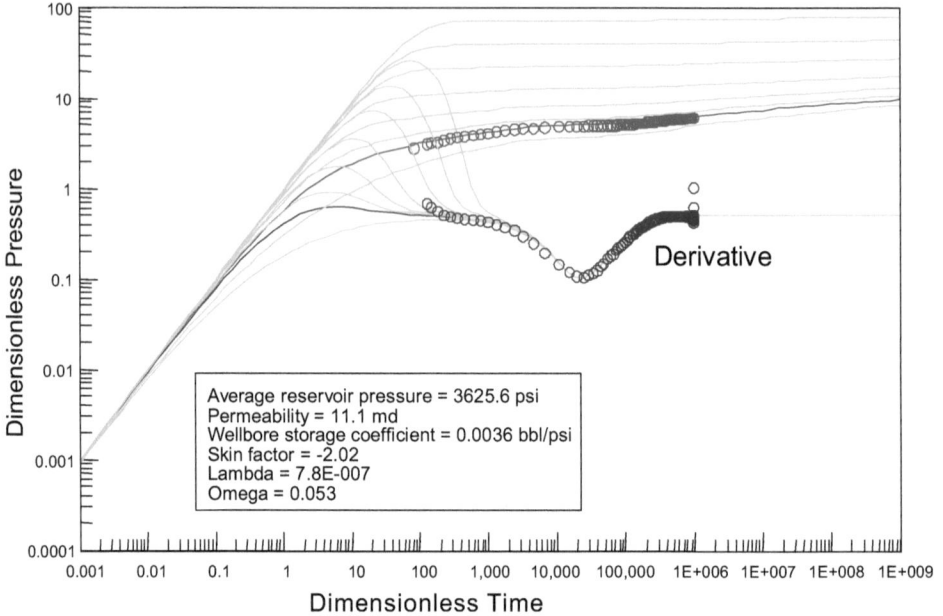

Fig. 3.19—Typical pressure-derivative plot in a naturally fractured reservoir during pseudo-semisteady-state.

Because fracture sets have a strongly preferred orientation, one of the primary objectives for interference tests in NFRs is to obtain an in-situ measure of reservoir permeability anisotropy; hence, more than two wells are commonly used, with two or more observation wells offset from the signal well in different directions. In-situ permeability can be computed between each pair of signal-observation well pairs (Kamal 1983). Multiple interference tests may be desirable from different wells to judge reservoir heterogeneity, which is usually high in fractured reservoirs.

Operationally, a source well is usually shut in while monitor wells instrumented with pressure gauges detect the response, both in time and in pressure. The pulse anisotropy can be quite high, with time delays of minutes between wells offset along the direction of an open fracture set, compared with hours or days for wells offset perpendicular to the fractures.

Tracer tests are an unambiguous method for determining direct connectivity between wells. A unique tracer is injected into a well, and surrounding wells are monitored for presence of the tracer. Yose et al. (2001) illustrate the coincidence in orientation of fractures measured in wells and well pairs showing connection with rapid tracer movement, injector-producer response interpretations, and well-performance-based interpretations at Norman Wells field (**Fig. 3.20**). The overall trend of hydraulic connectivity is similar to the predominant fracture trend.

3.8 Synthesizing All the Data

Numerous options have been discussed in this chapter for detecting fractures or their effect on well production. No approach is perfect, and the best approach for any particular field

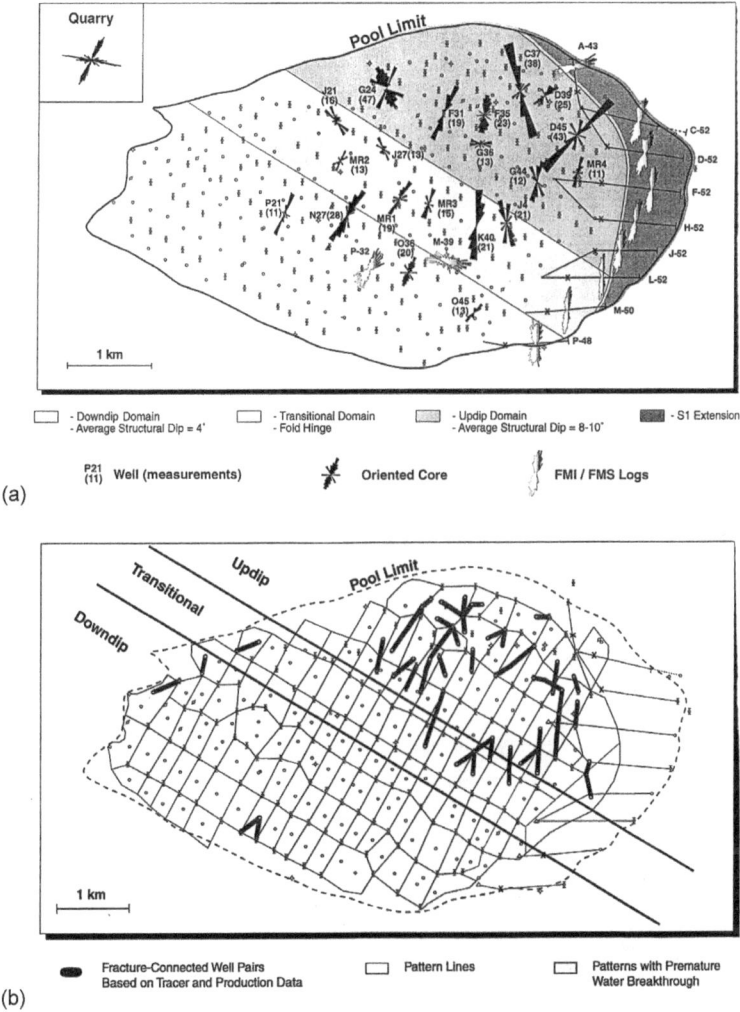

(a)

(b)

Fig. 3.20—Fracture trends from oriented core and image logs (a) and fracture-connected well pairs based on tracer and production data (b) at Norman Wells field, NWT, Canada, from Yose et al. (2001).

will depend on what data are available, the character of the field, the objectives of the characterization, and the tools available. Synthesis of multiple data sources generally improves understanding.

If we assume that a fieldwide fracture-density model is the objective, then well-test information can be used to anchor the model to the actual field permeability. We focus on the fieldwide fracture-density model because it carries fracture-set-specific information that is useful for more sophisticated simulation models (tensor-based effective permeability models, dual-porosity models, and discrete fracture network models).

Because of the heterogeneous nature of fracture systems, a perfect correlation between fracture density and productivity should not be expected, nor should it be forced on the data. A well can intersect a single spectacularly productive fracture in a region of low fracture density, giving a strong dynamic response. Likewise, a well in an area of high fracture density can fail to intersect any significant effective fractures. Relying solely on well-test results to steer the fracture-density model would ignore the stochastic nature of the well-fracture intersections. Values of fracture density are probabilistic estimates, and well-test results are effectively a roll of the dice within this probabilistic system.

The effective permeability of specific regions of an NFR may be known from well-test results. Fracture density has been interpolated over the same area from individual-well fracture data. The calibration points between these distinct data can be used to define a statistical correlation such that the fracture-density model can be transformed into an effective permeability for conditioning a full-field model.

DFN models can help in understanding the linkage between realistic fracture systems and well-test results (Dershowitz et al. 2000; Cacas et al. 2001). With this forward-modeling approach, realistic fracture system models are fabricated based on descriptions of fracture geometry and density (**Fig. 3.21**), and flow simulations through this DFN can be matched with the well test. Most DFN models have limited ability to model transfer of fluid from matrix to fractures; therefore, their usefulness is limited mainly to issues of fracture connectivity and pressure response relating to drainage from the fracture system for well-test purposes, not full-field simulation. Currently, dual-porosity/dual-permeability DFN modeling is an area of active research.

Finalizing the calibration may require changes of the model fracture system parameters locally or field wide. There is high uncertainty associated with fracture system descriptions—particularly with respect to fracture length. Altering a fracture-density model to calibrate with well-test results allows the interpreter to cycle back through the fundamental geologic data to verify whether proposed adjustments can be rationalized. Such pre-history-match modification allows a reservoir model to evolve within the constraints of all the data, thereby creating a reality-based NFR model. This model will be a robust basis for flow-simulation history matching and will be usable in future simulation models with minimal reworking.

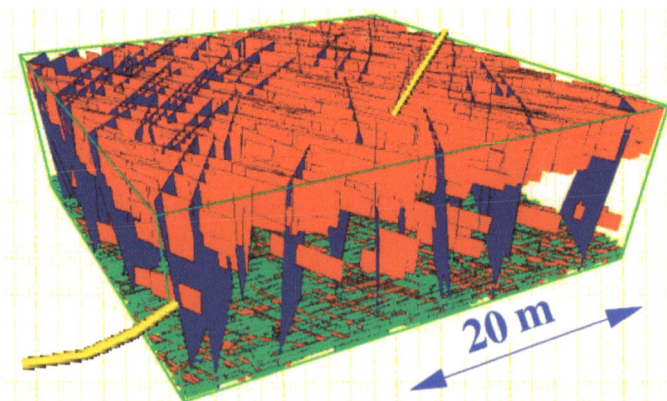

Fig. 3.21—Example of a DFN model (from Cacas et al. 2001). Used with permission of the Geological Society.

Chapter 4

Fluid Flow in NFRs

4.1 Introduction
The presence of fractures in a reservoir induces fluid flow much different from that of more conventional, unfractured reservoirs. Fluid displacement in a fracture network occurs because of the fracture system's higher conductivity compared to that of the matrix. As a fluid is injected into the fracture system, an exchange or transfer of fluids occurs between matrix and fracture system.

Fractures produce a variety of effects on fluid mobility in a reservoir that must be addressed in order to understand and predict reservoir production behavior. Most significant among fracture effects are:

- Matrix/fracture interactions, including matrix drainage and sweep.
- Permeability enhancement, both horizontal and vertical.
- Permeability anisotropy, horizontally and/or parallel to bedding (X-Y plane).
- Rapid fluid breakthrough or bypassing owing to flow directly within well-connected fractures.

We overview these issues here, beginning with issues of rock-fracture fluid transfer, and discuss them in the context of both their physical process and flow simulation.

4.2 Fracture/Matrix Interaction
4.2.1 Capillary Imbibition. The fluid transfer process during water injection is governed by the flow of water under an imposed pressure gradient (viscous force) and the spontaneous movement of water into the matrix, owing to capillary forces (imbibition). Spontaneous capillary imbibition occurs when the fractures contain a wetting phase and the resident oil in the matrix is the nonwetting fluid. During spontaneous imbibition, water imbibes into the matrix, and oil is expelled from the matrix to the fracture via a countercurrent mechanism. The rate at which water in a fracture may transfer to the rock can vary widely according primarily to the wettability of the rock matrix, the matrix permeability, and the intensity of fracturing. The "transfer function" that describes the rate of transfer and ultimate recovery of the nonwetting phase has been the subject of considerable study. The transfer function is directly incorporated into simulation models and thus describes the ultimate efficiency of water injection in a water-wet, naturally fractured reservoir. Con-

versely, if gas is injected into a fracture system, the nonwetting gas phase may be transferred into the rock matrix, thereby displacing the wetting oil phase via gravity drainage.

High-permeability fractures contain a small fraction of the total pore volume, and low-permeability matrix rock contains a significant portion of the reserves of a field. This contrast between reservoirs with high-permeability conduits and low-permeability matrix rock leads to challenging time-scale characterization problems because the wetting fluid in fractures displaces nonwetting phases in the matrix. This imbibition process is represented in many simulations as "dual porosity."

Some of the early work presented by Mattax and Kyte (1962) demonstrated that the rate of imbibition of water injected into a fractured reservoir with water-wet rock would occur according to the dimensionless rate parameter:

$$t_D = \sqrt{\frac{k}{\phi} \frac{\sigma \cos\theta}{\mu_w L^2}} . \quad\dots\dots\dots\dots\dots\dots\dots\dots\dots\dots\dots\dots\dots\dots\dots\dots (4.1)$$

To apply the experimental data to field-scale imbibition waterflooding using this mathematical model, dimensionless time (t_D), initially proposed by Mattax and Kyte, was modified by Ma et al. (1995).

$$t_D = t\sqrt{\frac{k}{\phi} \frac{\sigma \cos\theta}{\mu_g L_c^2}} , \quad\dots\dots\dots\dots\dots\dots\dots\dots\dots\dots\dots\dots\dots\dots\dots (4.2)$$

where t is imbibition time, k is permeability, ϕ is porosity, σ is interfacial tension, μ_g is the geometric mean of water and oil viscosities, L_c is the characteristic length defined by Ma et al. and θ is contact angle. The characteristic length for a matrix block in the reservoir is half the fracture spacing. Thus, for a given rock type, oil-to-water viscosity ratio and characteristic length or fracture spacing, laboratory experiments on small core samples could be scaled to larger block sizes representative of reservoir dimensions. Thus, laboratory recovery correlates to the inverse of block size (spacing). Eq. 4.2 demonstrates the importance of petrophysical characterization of not only the matrix (k and ϕ), but the fracture spacing (L_c) or the distance between fracture faces. The dimensionless time required to reach maximum recovery is inversely proportional to the square of fracture spacing, indicating that intensely fractured reservoir will rapidly reach maximum recovery, whereas lower fracture density implies longer times to attain maximum recovery.

The imbibition scaling equation is dependent on knowledge of wettability (or capillary pressure) of a given rock via the $\sigma \cos\theta$ term. Wettability is usually characterized by various wettability indices such as the Amott index. Values of wettability index and imbibition capillary pressure allow determination of endpoints expected during water injection into a fractured system. High wettability indices near +1 indicate strongly water-wet rock and significant imbibition potential. Low index numbers indicate weakly water-wet systems with limited ability to imbibe, and negative wettability indices refer to oil-wet rock. There are no guidelines regarding the wettability index of a particular lithology. Carbonates are generally expected to be oil-wet, and sandstones are expected to be water-wet, although there are exceptions to both cases. Wettability should be evaluated on a case-by-case basis for each reservoir, using great care to properly perform the laboratory evaluation.

To understand the imbibition response of a reservoir, two critical factors need to be addressed: determining the correct wettability of the reservoir rock, which is a laboratory-scale phenomenon, and accurately determining the water-rock contact area (fracture surface area), which is a field-scale phenomenon. Wettability can be determined from simple experiments of spontaneous imbibition and forced imbibition. Proper sample (core) preservation plus proper cleaning and handling are critical if these experiments are to provide accurate results. The water-rock contact area is a function of the fracture density; however, the surface areas used in simulation typically are not equal to those obtained directly from static borehole characterization or well-test analysis. Effective fracture surface area is a first-order uncertainty for simulation, and sensitivity to it should be evaluated over a large range of values spanning several orders of magnitude.

4.2.2 Matrix Capillary Pressure for Imbibition. Use of an appropriate capillary pressure function is necessary as input into a dual-porosity simulation (not a drainage capillary pressure). Proper measurement of imbibition capillary pressure is a difficult task, with results that are predicated on the wettability of the sample. Sufficient exposure of the rock surface to crude oil is necessary to establish the correct wettability and thus determine an accurate imbibition capillary pressure. Since this is the driving force for the transfer mechanism during imbibition, the importance of this parameter cannot be discounted.

4.2.3 Gas-Gravity Drainage. If gas is injected into a fracture system, the nonwetting gas phase may be transferred into the rock matrix, thereby displacing the wetting-oil phase via diffusion or gravity drainage. Gravity drainage is dependent on the height of matrix blocks, the density contrast between the wetting and nonwetting phases, and the interfacial tension between the wetting and nonwetting phases. Important applications of gravity drainage may be found in the Yates field, Cantarell, and Middle Eastern fields such as Haft-Kel. Many fractured fields are excellent candidates for gas injection.

Gas-gravity drainage has proven to be an important transfer mechanism as gas is injected into the fracture system. The rate of oil recovery is a direct function of permeability of the matrix. If the gravitational head can overcome the capillary restraining forces, gravity drainage of matrix blocks is a feasible recovery mechanism. The balance between capillary and gravity forces determines (1) whether a matrix block will drain under gravity forces and (2) the residual oil saturation in matrix blocks. The ratio of capillary to gravity forces (CGR) for a single matrix block is given by:

$$\text{CGR} = \frac{P_c}{P_g} = \sqrt{\frac{k}{\phi}} \frac{\sigma \cos\theta}{\Delta\rho g h_f}, \quad \dots\dots\dots\dots\dots\dots\dots\dots\dots\dots\dots\dots\dots\dots (4.3)$$

where $\sigma \cos\theta$ is the capillary pressure term that accounts for the capillary pressure of the porous system and the wettability of the rock surface. The gravity term opposing the drainage of the denser hydrocarbon through the rock matrix is $\Delta\rho g h_f$. A ratio of these two forces determines key factors that initiate gas-gravity drainage. The interfacial tension determines the magnitude of capillary retention in the matrix block. The vertical height of the fracture (h_f) allows the gravity head to overcome restraining capillary forces. Schechter et al. (1994) show that a ratio of $P_c/P_g < 1$ results in initiation of gravity drainage from the matrix block. **Fig. 4.1** shows the interrelationship between the parameters. If the $P_c/P_g = 1$,

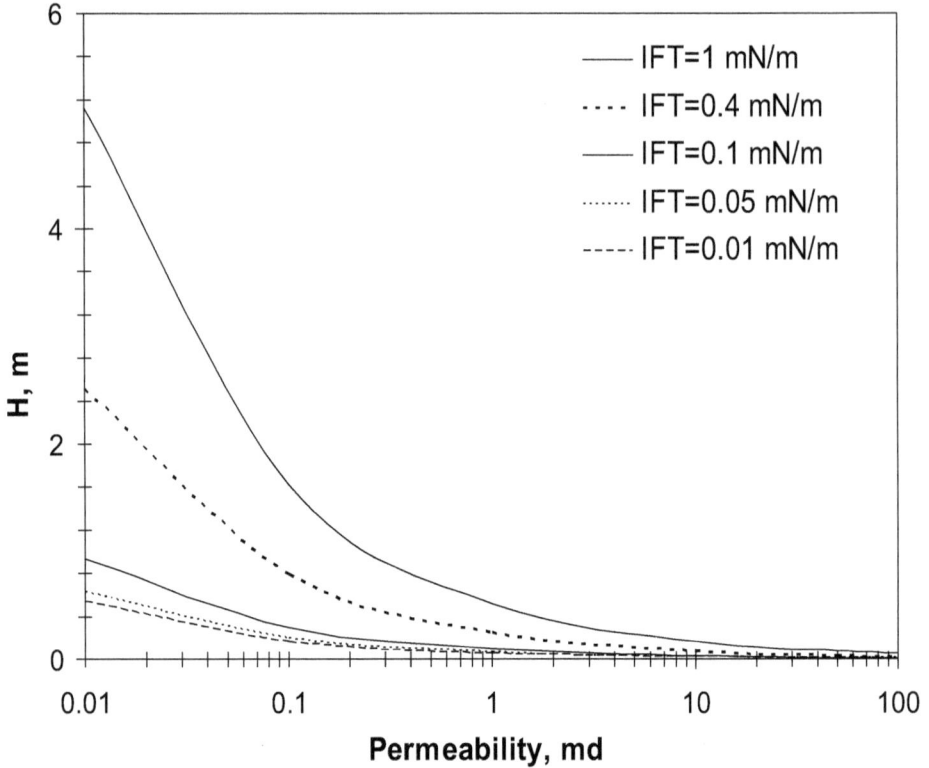

Fig. 4.1—The height of a matrix block (or fracture) vs. permeability of the matrix rock to initiate gas-gravity drainage. Different curves represent different values of IFT, of which $P_c/P_g = 1$.

as is shown in Fig. 4.1, modest block heights combined with low interfacial tension will initiate gas-gravity drainage.

4.3 Yates Field

The primary reservoir of the Yates field, in the Permian Basin of west Texas, is a well-studied gravity-drainage reservoir. This massive carbonate reservoir, with reserves of more than 1.3 billion bbl, is similar to some large naturally fractured carbonate reservoirs in the Middle East. According to Craig (1988), the main reservoir at Yates is the Permian-age San Andreas formation, which is up to 229 m thick. Throughout a significant portion of this reservoir, the fractures underwent solution enlargement, leading to cavern development and extremely efficient connectivity. The reservoir is in pressure communication across its entire thickness, and this pressure continuity extends more than 30 m into less-productive overlying formations.

A history of the Yates field is summarized by the following:

1. The field lies at a shallow depth of approximately 500 m. Initial reservoir pressure was about 700 psig, and is approximately on a hydrostatic gradient from the surface. Its temperature was 28°C.

2. The San Andreas reservoir did not have an initial gas cap but developed a secondary gas cap as a result of primary depletion, aided by the high connectivity from fractures and caverns. The expanding secondary gas cap helped to drive oil recovery for several decades.
3. The field has had a strong waterdrive, but the simultaneous movement of the gas/oil contact from above and the water/oil contact from below have been balanced for efficient oil recovery.
4. Gravity drainage has been utilized as an important element in secondary oil recovery. In the high-relief eastern portion of the field, CO_2, N_2, flue gas, and steam were all injected into the gas cap. In the flatter western area, a low-tension waterflood was applied.
5. After drilling a large number (> 900) of vertical wells, it was realized that many of these were unnecessary because of the high fracture connectivity in the reservoir.
6. A substantial number of successful horizontal wells have been drilled throughout the field.

4.3.1 Relative Permeability in the Matrix and Fractures. Two-phase relative permeabilities are strongly affected by capillary forces, both in the fracture and the matrix. The matrix relative permeability is measured in the laboratory either in the cocurrent or countercurrent modes for imbibition. Concurrent imbibition implies that the displacing and displaced fluids move in the same direction; however, one can envision a countercurrent mechanism in which water is imbibed into the matrix toward the center of the matrix block and oil is displaced in the opposite direction toward the fracture. Laboratory work (Bourbiaux and Kalaydjian 1990) has demonstrated that countercurrent relative permeability is less significant than concurrent relative permeability in similar rocks. Relative permeability in fractures is assumed to show straight-line functions with endpoints at zero and 100% saturation, since fractures are thought to behave as open conduits or "pipes." Although this is intuitively incorrect, especially where apertures of the fractures result in nonzero capillary pressure, there is no generally accepted method to vary relative permeability within the fracture system.

4.3.2 Transfer Function. Dual-porosity simulations require an expression for rate of fluid transfer between matrix rock and fractures. The transfer of fluids is determined by capillary imbibition or gravity drainage. The transfer function utilizes the shape factor to determine rates of imbibition during water injection into a fractured reservoir, thus effectively combining fracture characterization (fracture spacing) and the rate of imbibition as determined by wettability, relative permeability, and capillary pressure.

The transfer function is defined as the transfer of fluids from the matrix to the fracture, with an appropriate geometric factor that accounts for the characteristic length and the flow area between the matrix and the fracture. The fluid transfer rate per unit volume of rock is calculated from the following expression:

$$q = \frac{\sigma_f K_m}{\mu} (p_m - p_f). \quad\dots\dots\dots\dots\dots\dots\dots\dots\dots\dots\dots (4.4)$$

The transfer of fluids from matrix blocks into the fractures is assumed to be in a steady-state condition and is a function of viscosity of the fluid, pressure drop between the matrix and fracture systems, and matrix-rock properties related to geometry and eventually interconnectivity of the matrix block. The shape factor, σ_f (the subscript f is unconventional but is specified here to distinguish the shape factor from interfacial tension, which also uses σ as its conventional symbol), defines block geometry, although the assumption of cubic matrix blocks is inherent in such models.

4.3.3 Geometry of Fracture Network—Shape Factor. The shape factor describes, for simulation, both the surface contact area between fractures and matrix and an idealization of the geometry of the matrix blocks. Dual-porosity models rely on the shape factor to determine parameters like the total permeability of the fracture network and the tendency for fluids to move in a direction with preferential permeability.

The shape factor is generally defined as:

$$\sigma_f = 4\left[\frac{1}{L_x^2} + \frac{1}{L_y^2} + \frac{1}{L_z^2}\right]; \quad \dots\dots\dots\dots\dots\dots\dots\dots\dots\dots\dots\dots\dots (4.5)$$

L_x, L_y, and L_z refer to the matrix block dimensions in each of three mutually perpendicular directions, often referred to as a sugar-cube arrangement.

The shape factor σ_f is an idealized mathematical representation of the fracture system in the reservoir, representing fracture surface area as well as spacing. The value of L may relate to the relative spacing of fractures in one direction in the reservoir, but generally it will not equal the true geological average fracture spacing. The spacing of fractures can be highly variable, such that some matrix blocks will be very small and some quite large; however, the value of L in any direction in a simulation cell is fixed. Furthermore, natural fractures do not always interconnect, and they are generally not orthogonal, both of which are factors that influence the distinction between the value of L and the natural fracture spacing. Horizontal effective fractures are extremely rare in nature, and should be included in a simulation model only if a compelling reason exists.

Obviously, as fracture density varies across a field, the shape factor will also vary. This requires substantial understanding of determination of shape factor by means of the well-testing procedure outlined previously. In addition, the shape factor should be indicative of the inflow performance of the reservoir and should agree with well-test data as well as core data if available. For instance, the value of shape factor in an intensely fractured area will be relatively small. The well should have greater productivity indices compared to wells completed in less intensely fractured zones. Ideally, core descriptions from the two wells will indicate a more intensely fractured rock compared to core from the well that exhibits lower PIs. Verification of this alteration in shape factor across a reservoir by well testing, well performance, and core data can be used in development of a static reservoir model for reservoir simulation.

In many situations, isolated large fractures will produce high well PIs even if the bulk fracture density is low, thereby confounding any relationship between fracture density and well PI. This is one pitfall of the dual-porosity idealization of linking well PI to fracture density.

4.3.4 Connectivity of Fracture Network. One elusive parameter that has a large effect on dual-porosity simulation is connectivity of the fracture network. Even parallel fractures in a regional fracture setting exhibit some degree of directional variability. Consequently, closely spaced fractures sometimes intersect, or become connected by short cross fractures.

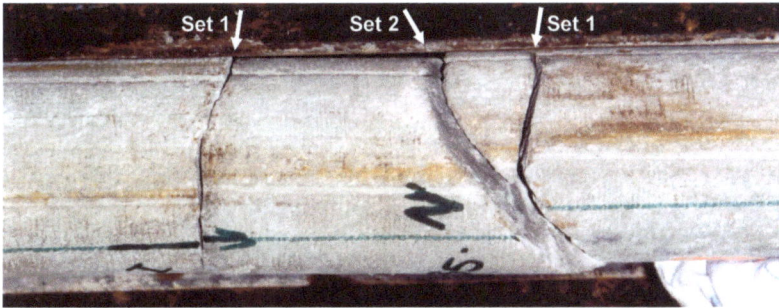

Fig. 4.2—Two natural fracture sets intersecting in horizontal core.

The result of fracture intersections is reduction of anisotropy. Networks with multiple well-developed fracture sets, such as are observed in the Spraberry trend (**Fig. 4.2**), are characterized by significant connectivity. However, the high flow anisotropy of the Spraberry trend suggests that crossing fractures are significant only locally (Figs. 1.7, 1.10, and 5.9).

4.4 Simulation of NFRs

Production of NFRs typically proceeds according to one or more of the following schemes:

- Primary depletion of the fracture system as the fractures unload stored hydrocarbons, sometimes leading to very high IPs.
- Solution gas drive, with the critical gas saturation being a crucial factor in ultimate recovery efficiency.
- Water influx in the case of an aquifer situation, sometimes leading to irregular water breakthrough problems, with the wettability of the rock being a primary determinant of the extent of displacement efficiency of the reservoir rock as the water/oil contact rises into the oil zone via the fractures.
- Water injection into the aquifer or into a horizontal layer of fractured rock to displace more oil by means of capillary imbibition, and injection to maintain reservoir pressure or into underlying aquifers to mobilize the oil water contact at a greater velocity. Once again, the keys to these processes are the wettability of the reservoir rock and the intensity of the fracture network.
- Gas injection into crestal wells to sweep oil downward on an anticline/steeply dipping reservoir or injection into a horizontally layered system with vertical fractures in the pay zone. The primary mechanism encountered as the gas moves through the fractures and surrounds matrix blocks is gravity drainage.

There are a variety of methods used to simulate these scenarios. A classic single-porosity simulation uses a single set of data for both the porosity and permeability of the hydrocarbon reservoir. Dual-porosity models have been developed in the last few decades, based on the initial work of Warren and Root (1963). Their work recognized that fractures tend to provide a majority of the reservoir permeability, whereas the matrix rock contains most of the stored hydrocarbons. A common approach has been to separate the fractures and matrix into discrete continua in which each element has a unique value of both permeability and porosity. The dual-porosity model also allows fluid exchange or the transfer mechanism that determines the rate of hydrocarbon flow from the matrix rock to the fracture network to be quantified. One negative aspect of the dual-porosity approach is the requirement of more than twice the data that is needed for a single-porosity simulation. Often, the data required for a dual-porosity simulation are poorly characterized, thus resulting in erroneous or poorly constrained model input. Furthermore, the computational demands of dual-porosity models are much higher than single-porosity models: run times for dual-porosity simulation take about 5 times longer than a comparable single-porosity model.

Further complicating the situation is the fact that many reservoirs display dual-porosity behavior, yet the configuration of the fractures may be vastly different from one reservoir to the next. For instance, one reservoir may have horizontal bedding with low-permeability matrix rock and a series of regional fractures. Another fractured reservoir may be in an anticline with significant bedding dip and the primary hydrocarbon storage in vuggy porosity, with microfractures interconnecting the vuggy porosity. Even though the fracture systems are very different in the two situations, they both are affected by the capillary pressure, gravity drainage, and fracture geometry.

Throughgoing regional fractures may greatly enhance the permeability of the reservoir, with the matrix permeability and fracture spacing acting to control the rate of fluid transfer. Alternatively, in vuggy zones with the vuggy porosity interconnected by microfractures, the interconnectedness of the microfractures may be the limiting rate factor for fluid transfer. Both of these examples represent dual-porosity behavior, yet each has distinct and unique fracture characteristics. The variety of naturally fractured reservoirs encountered in nature indicates the challenge involved in developing simulation tools that represent the myriad of physical behaviors encountered in NFRs.

Flow-simulation approaches currently available for modeling NFRs are summarized below:

Dual-Porosity (and Dual-Permeability/Dual-Porosity) Models (DKDP).

1. Dual-porosity models allow bulk fluid movement only through the fracture system (matrix-to-fracture transfer, then flow along fractures), whereas dual-permeability/dual-porosity models allow flow from matrix block to matrix block as well as fracture-to-matrix and matrix-to-fracture flow.

2. The DKDP model allows highly adjustable matrix/fracture transfer physics and surface conditions, which are important for some reservoirs.

3. Fractures are not modeled discretely, but rather are treated as an aggregate flow regime distinct from the matrix flow regime.

4. These models are complex, with a large number of poorly known variables. Their computer run times can be considerable.

5. DKDP models are most appropriate where the fracture system is well connected and where complex matrix/fracture transfer physics are important (e.g., imbibition and gravity drainage).

Single Effective Medium.

1. The flow properties of fractures and matrix are combined to provide a single effective permeability description for grid cells. Models range from simple permeability multipliers to more complex permeability tensors that seek to model permeability anisotropy (Oda 1985; Lee et al. 2000).

2. Fracture/matrix fluid interaction is addressed by using pseudorelative permeability functions.

3. The single effective medium approach is most appropriate where the fracture system is highly variable in its development, where reservoir performance is governed more by fluid movement rather than matrix/fracture interactions, and for single-phase flow or where pseudorelative permeability suffices to describe multiphase effects.

DFNs.

1. In a DFN model, the geometry of individual fractures is fully defined in a reservoir, and the flow of fluids through them is modeled. This is the most geologically realistic of present modeling approaches for tracking the movement of fluid within a fracture system (see Chapter 3).

2. DFN models do not presently allow integrated transfer of fluids between fractures and matrix blocks, so their applicability for full-field simulation is limited. However, they are useful as a basis for calibrating fracture effects on individual well behaviors. DFN modeling can be used to evaluate fracture-system scenarios based on well-test analysis, particularly the connectivity character of fracture systems using single or multiple well-production tests.

The alternative simulation methodologies available reflect the variability in NFRs. No modeling approach is optimal for every situation, so the choice should be made according to the character of the reservoir and the priorities of the model objectives. Choices are not clear-cut; there is considerable debate among specialists regarding the applicability of different modeling approaches. Ongoing research is focused on development of modeling approaches that are hybrids of existing model technologies, to take advantage of the strengths of each model approach.

Chapter 5

Case Histories of NFRs

5.1 Midale Field

A CO_2 flood of the Weyburn formation, Midale field, in Saskatchewan, Canada, demonstrates the significant effect fractures can have on the productivity of a reservoir (Beliveau et al. 1993; Elsayed et al. 1993; Baker et al. 2000). The field was discovered in 1953 and developed on 80-acre spacing. In 1962, the field was unitized for water injection. The anisotropic nature of the reservoir was apparent from water-cut maps (**Fig. 5.1**). Wells were originally oriented in an inverted nine-spot, so for each injector well, a corresponding on-trend (fracture-parallel), off-trend (fracture-normal), and diagonal (oblique to fractures) well existed.

The Weyburn formation reservoir is an intensely fractured, 40-foot-thick vuggy limestone (the vuggy zone) with porosity of 15% to 20% overlain by a less intensely fractured 30-foot-thick chalky dolomite (the marly zone) having porosity of 20% to 35% (**Fig. 5.2 and Table 5.1**). The vuggy zone contains rock of two facies—shoal and intershoal—that have similar porosity but show different fracture spacing.

Cores from vertical and horizontal wells and image logs show that fractures are abundant. Fractures are vertical or near-vertical (perpendicular to bedding) and are generally open. They range in height from several inches to about 15 feet. The fractures range from small cracks that can be fitted back together to wider-aperture features with solution-enlarged gaps. Fractures strike almost exclusively NE/SW, which is evident in the flow anisotropy of reservoir fluids. Fracture spacing (horizontal distance between fractures) is about 2 to 3 feet in the marly zone, 2 to 4 feet in the intershoal vuggy zone, and 1 foot in the shoal vuggy zone.

After water injection had established residual oil saturation, a pilot CO_2 flood was conducted from 1984–1989 on 4.4 acres. The flood pattern consisted of four injection wells surrounding three production wells (**Fig. 5.3**). The pilot project consisted of a multiwell interference pressure transient and a salt tracer study. Cased-hole fiberglass wells were used for time-lapse monitoring of saturation within the pattern. Multidirectional pulse tests using sensitive downhole quartz gauges showed that wells located along the fracture trend responded quickly and with the largest change in observed pressure magnitude. Wells offset diagonal to the fracture trend responded with less magnitude and with a slower peak arrival. The response of wells offset perpendicular to the fracture trend was hardly perceptible. The flow anisotropy varies from 2:1 to 150:1, averaging 25:1. The matrix properties are rela-

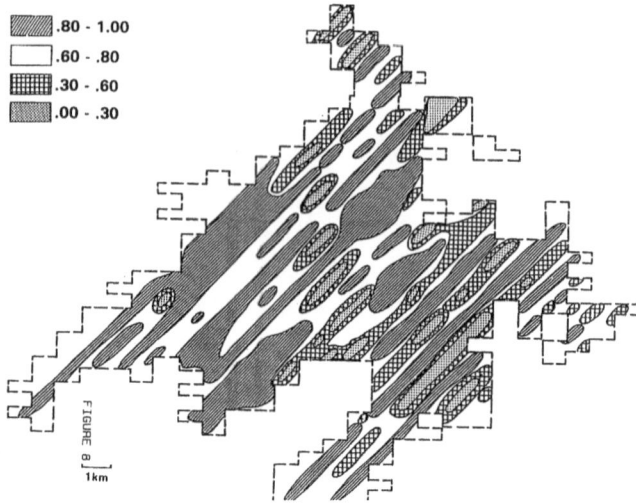

Fig. 5.1—Midale unit water-cut distribution (after Beliveau et al. 1993).

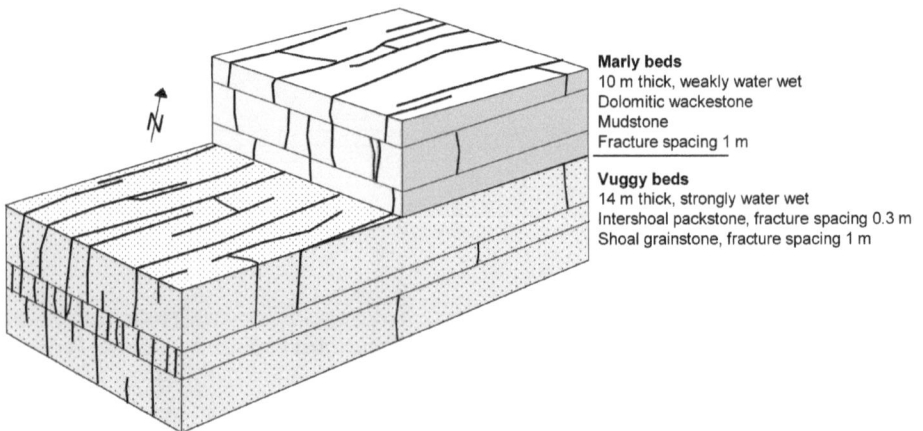

Marly beds
10 m thick, weakly water wet
Dolomitic wackestone
Mudstone
Fracture spacing 1 m

Vuggy beds
14 m thick, strongly water wet
Intershoal packstone, fracture spacing 0.3 m
Shoal grainstone, fracture spacing 1 m

Fig. 5.2—Schematic illustration of the Weyburn formation reservoir at Midale field.

tively constant throughout this test area, so the flow anisotropy is mainly a function of fracture system variability. Salt tests and radioactive tracer tests corroborated the rapid communication along the NE/SW fracture orientation, with some tracers taking less than 2 hours to travel 200 feet.

Despite the rapid movement of injected water along the fracture trend, injected CO_2 took nearly a month to move along the fractures and break through in producing wells, indicating that CO_2 matrix/fracture interaction involved different mechanisms from water matrix/fracture interaction. Likewise, oil production dropped to zero during the first month of CO_2 injection, which contrasted with the low but steady oil cut during water injection. However, after 5 weeks of CO_2 flooding, significant incremental oil recovery was observed.

TABLE 5.1—WEYBURN RESERVOIR PARAMETERS	
Lithology	Limestone/Dolomite
Depth at reservoir midpoint, ft	4,600
Area, acres	~26,000
Net pay, ft	60–80
Average porosity, %	20–35
Matrix permeability range, md	1 to 10
Maximum well-test permeability, md	200
Average reservoir pressure, psi	2,600
Minimum miscibility pressure, psi	2,250
Water saturation, %	23.6
Initial GOR, scf/STB	1530
Initial formation volume factor (FVF), RB/STB	1.78
Oil gravity, °API	32

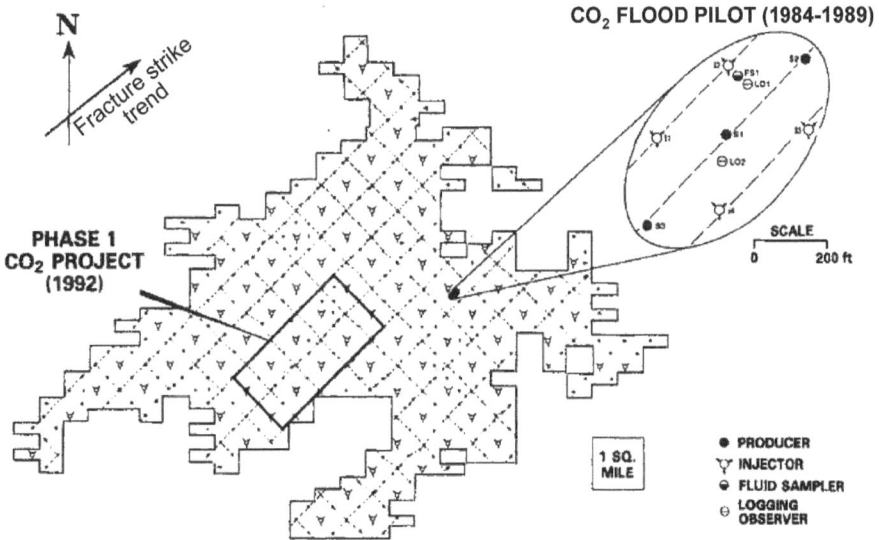

Fig. 5.3—Midale field, showing CO_2 flood pilot and project locations (from Beliveau et al. 1993).

This indicated that viscous stripping of oil from the vicinity of fractures was the recovery mechanism of the waterflood, whereas the CO_2 was penetrating the matrix and recovering incremental tertiary oil by gravity drainage.

The logging program showed that gravity played a major role in recovery. CO_2 injection was only into the vuggy zone, but the response at observation wells was mainly in the marly zone, indicating that CO_2 was able to move vertically across the reservoir.

A favorable condition exists in Midale, where chalky dolomites of the marly zone that are weakly water-wet are underlain by strongly water-wet vuggy limestone. Injected water tends to move downward in the fracture system aided by the water-wet limestone, whereas CO_2 rises in the vertical fractures to the marly zone, which was weakly water-wet and thus had higher residual oil saturation.

Midale field was simulated by two independent groups who used different approaches, and both report satisfactory results. Beliveau et al. (1993) used a standard dual-porosity formulation to address the mechanisms of drainage, imbibition, and local pressure equilibrium, whereas Elsayed et al. (1993) used a single-porosity model with strong flow anisotropy. In the dual-porosity model, fracture geometry was addressed with a dominant set oriented parallel to the observed natural fracture trend, plus a set of special connections perpendicular to the dominant trend to capture the uncommon effects of off-trend fractures. For the Midale reservoir, low fracture conductivity in the direction perpendicular to the primary fracture orientation resulted in a significant pressure drop. Pressure drop of this magnitude may result in a viscous drive, with injected fluids displacing oil from the matrix. Thus, fluid could move off trend either through matrix permeability or off-trend fractures.

A modified dual-porosity simulation was used in Midale to describe the primary physical mechanisms of importance in this naturally fractured reservoir. For water injection, the wettability and capillary imbibition play a key role. Gas injection is dominated by multi-contact miscibility, resulting in lowered interfacial tension and subsequent gravity drainage. In the interwell area, pressure drop in line-drive patterns betweens rows of injectors and rows of producers may result in a viscous pressure drop.

In Midale, the history-matching procedure allowed that the vuggy limestone at the base of the structure was strongly water-wet. Injection wells placed parallel along with production wells produced no oil during water injection if the wettability index was set to neutral. However, a significant response was seen along the fracture trend if the simulation used the actual wettability index (or capillary pressure) from special core analysis. Because the pressure drop along the fracture network was low, oil recovery along the high-permeability orientation was a clear indication of capillary imbibition.

In their distinctly different single-porosity flow-simulation model, Elsayed et al. (1993) used an average 6-to-1 permeability anisotropy to address the on-trend to off-trend fracture flow effect. They defined the anisotropy based on observed fractures and on results obtained from production and numerous drillstem tests. They decided to use a single-porosity model after recognizing that differences in saturation and pressure between the fractures and matrix are small. The lower computing demands of a single-porosity simulation allowed them to use smaller grids and thus include more geologic detail compared with a dual-porosity model. They report that a history match was achieved with minimal changes to the basic input parameters from the reservoir characterization model.

The achievement of satisfactory history-match results from both dual- and single-porosity models for the same field shows that distinctly different simulation approaches can achieve similar results, demonstrating that there is no single "right" approach to simulation modeling of NFRs.

Horizontal injection and production wells have been utilized recently to further improve recovery in Midale. From **Fig. 5.4**, it is evident that horizontal producing wells have increased oil rates across the field by more than 50%.

The fracture-characterization and simulation effort at Midale field was used to optimize plans for a commercial-scale CO_2 flood project. The simulation model derived from the pilot CO2 flood indicates that an incremental tertiary recovery of about 20% of original oil in place (OOIP) can be expected. These pilot-flood results form the basis for field manage-

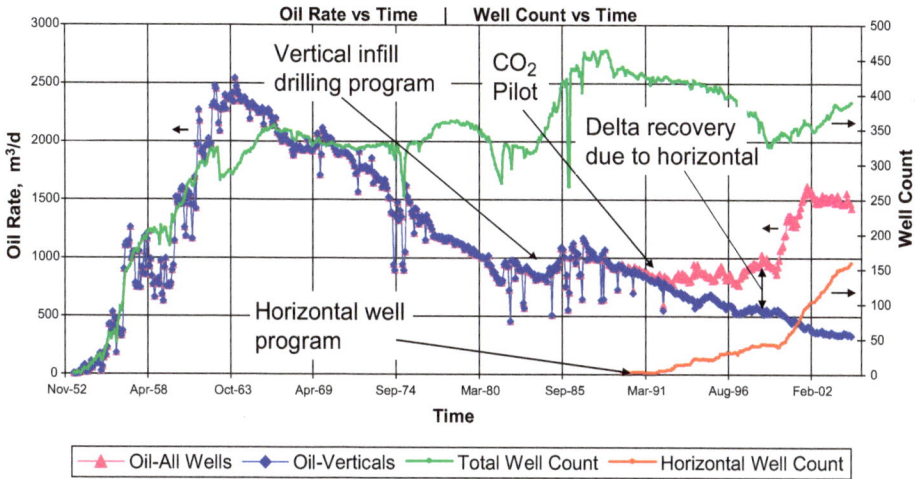

Fig. 5.4—Injection and production history for Midale field.

ment during the commercial project, including infill drilling with vertical and horizontal wells and injection patterns.

5.2 Spraberry Trend

The naturally fractured Spraberry trend area in west Texas, introduced in Chapter 1, was discovered in 1949 and continues to produce 60,000 B/D from more than 9,000 wells. The producing extent encompasses more than 2,500 square miles. Spraberry reservoirs originally contained some 10 billion bbl of oil (**Table 5.2**). In the 50-plus-year pumping history, 790 million bbl have been produced, representing a recovery factor of less than 10%. This case history describes some of the historical background of Spraberry waterflooding and reports recent results of water and CO_2 injection in the Spraberry E.T. O'Daniel unit.

TABLE 5.2—SPRABERRY RESERVOIR PARAMETERS	
Original reservoir pressure, psia	2,300
Saturation pressure, psia	1,840
Reservoir temperature, °F	138
Initial water saturation, %	35
Initial oil saturation, %	65
Matrix porosity, %	10.0
Effective permeability, md	2.0–183.0
Matrix permeability	0.1–1 md
Pore compressibility, psi^{-1}	4.00 E–6
Fracture porosity, %	0.1
Fracture permeability ratio (k_{xf}/k_{yf})	15/0.25
Estimated shape factor (σ)	1.47
Dominant fracture orientation	NE/SW

More than half a century since discovery, the reservoir has maintained status as one of the more complicated naturally fractured reservoirs. Understanding or predicting its response to water injection has proved difficult. Essentially, displacement and sweep efficiency are not well understood. To this day, there is no engineering consensus regarding the impact of water injection in Spraberry reservoirs. Most operators do not believe water injection is effective; therefore, a relatively small volume of water has been injected.

Originally, water and gas injection in fractured reservoirs were applied in a haphazard fashion. The presence of fractures would lead to rapid channeling and excessive production of injected fluid. However, Brownscombe and Dyes (1952a) performed imbibition experiments on Spraberry rock and identified this as a possible drive mechanism if injected water within the fractures contacts tight matrix rock, where a majority of the oil is stored. Water-flooding was initiated in the 1950s to test this idea. Two small pilots and limited water injection were pursued, yet not much enthusiasm was in evidence once initial and long-term results were analyzed.

Atlantic conducted an early water-injection test in the 1950s based on the results of imbibition experimental efforts presented by Brownscombe and Dyes (1952b). Enright (1954) describes the behavior of offset wells across a lease line that showed incremental oil recovery as a result of injection in three wells. However, the center test well showed no response, so the test was deemed a failure. This test established the N50°E orientation for the natural fractures. Barfield (1959) described an effort by Humble Oil Co. beginning in 1955 on an 80-acre pilot in the Midkiff area of Spraberry. Production in the center well of a 5-spot pattern increased from 70 to 250 bbl of oil per day. Humble determined the ratio of permeability *along* the fracture orientation to *perpendicular to* the fracture orientation to be 144:1, as is shown in Fig. 1.12. The response of this single well in the Humble pilot dictated future development of line-drive waterflood patterns in Spraberry.

A large water-injection project in the Driver unit in the early 1960s, described by Elkins and Skov (1960), failed to reproduce the results of the pilot conducted by Humble in the Midkiff unit. Water-injection wells were aligned parallel to the NE/SW fracture trend and perpendicular to a line of production wells. The general idea of line-drive patterns was to build interference between injectors along the fracture trend and force injected fluid to a line of producing wells, also aligned along the primary fracture trend. However, in the large-scale waterflood project in the Driver unit, no dramatic increase in production was noted in rows of production wells as was observed in the Humble pilot. Yet substantial increases in production were noted in several wells near the unit boundaries and some wells north of the unit. Unfortunately, historical documentation on well-by-well behavior is rarely available for Spraberry waterfloods. Fluid migration across lease boundaries has hampered assessment of water injection over the years.

Many of the early floods set the negative tone now associated with water injection in Spraberry. Upon closer analysis, however, it appears that early waterfloods deemed failures were actually successful at increasing reserves, perhaps unbeknownst to the operators. Production wells located on-trend from injection wells along the primary fracture axis have always responded favorably. Wells off-trend tend to show little response to water injection given sufficient distance between injector and producer, as is the case in early Spraberry floods. The one exception was the center well in the Humble pilot and a successful pilot in the O'Daniel unit reported by Guidroz (1967).

The Midale field injection behavior, described previously, presents a contrast with the behavior of Spraberry. At Midale, injected water was demonstrated to move along the primary fracture orientation of N50°E, and fluid eventually was forced perpendicular to the primary fracture trend through matrix permeability or cross fractures. Thus, sweep efficiency was sufficient to develop most of the field on 80-acre, inverted nine-spot patterns. In Spraberry, although fracture orientation and spacing are similar, the behavior of fluid recovery differs, and a significant effect on reservoir performance occurs.

5.2.1 Fracture Characterization in the O'Daniel Pilot. The E.T. O'Daniel Pilot is located in Section 4 T2S, Block 37, of the E.T. O'Daniel lease in Midland County, Texas. This site was selected due to high oil recovery that occurred from both primary and water injection here. The pilot configuration is shown in **Fig. 5.5**. Fifteen wells were completed with four CO_2 injection wells, three central production wells, and two logging observation wells. A horizontal-core well was obtained to study the fracture system (Lorenz et al. 2002). The well pattern is oriented along the major fracture trend with the three producers forming a line parallel to the primary fracture trend. Producers are flanked by four CO_2 injectors and surrounded by six water injectors in an approximately hexagonal pattern. The overall area confined by six water injectors is 67 acres, while the area enclosed by the four CO_2 injectors is 20 acres.

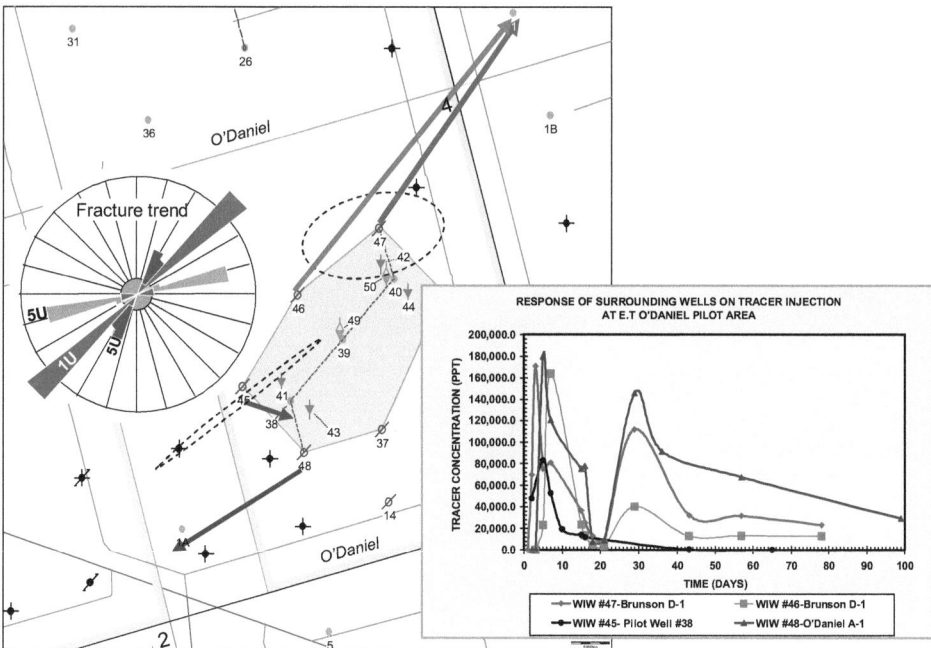

Fig. 5.5—Map of Spraberry pilot area. Rose diagram shows fracture trends in horizontal core, shown at the location of O'Daniel Well 28. Results of tracer flow show four primary response peaks. Responses are consistent with orientations of horizontal-core fractures. Response strength diminished from O'Daniel Well 46 to Brunson D Well 1, O'Daniel Well 47 to Brunson D Well 1, O'Daniel Well 48 to O'Daniel A Well 1, and O'Daniel Well 45 to O'Daniel Well 48. Pressure interference results are shown as dashed ellipses around injection wells 45 and 47.

5.2.2 Horizontal Core. The target zones for the pilot are two sand units separated by approximately 150 ft of nonreservoir rock. A single fracture set with a NE/SW orientation was expected. However, horizontal core from O'Daniel Well 28 showed three separate fracture sets within the two thin pay zones of the Upper Spraberry sand: one fracture set in the upper 1U sand (average orientation of N43°E) and two intersecting fracture sets (average orientation of N32°E and N70°E) in the lower 5U sand, as shown in **Fig. 5.6.** Fractures were not present in the underlying shale of each of the reservoir units; however, natural fractures were present in the shaley sands above the 1U and 5U zones **(Fig. 5.7)**. The recognition of three distinct fracture sets helps explain some of the puzzling behavior observed at Spraberry. Schechter et al. (1996) show that the three fracture sets clearly have implications concerning network connectivity, and many of the results of the Spraberry pilot demonstrate that one, two, or all three sets of fractures may have a regional influence on reservoir performance.

5.2.3 SRIT and Minifrac. A step-rate injection test (SRIT) and minifrac were performed in O'Daniel Well 37 to understand parting pressure and stress anisotropy. The pressure response from SRITs was dominated by fracture propagation and stress sensitivity. The SRIT and minifrac demonstrated two consequences as a result of the very low stress anisotropy measured between the shale and the reservoir rock: (1) hydraulic fractures are not contained

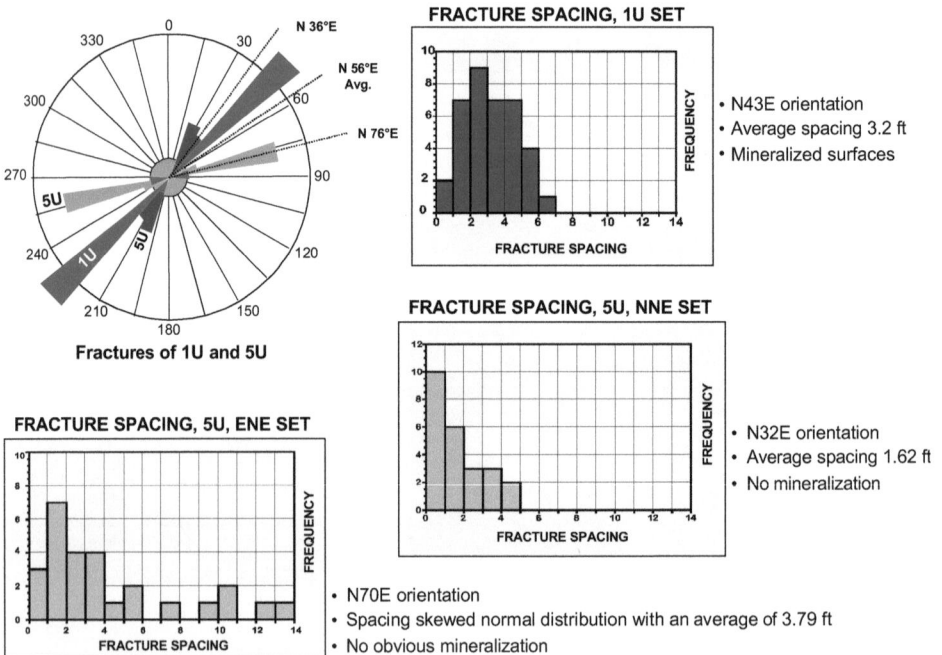

Fig. 5.6—Character of fractures in horizontal core from E.T. O'Daniel Well 28. Three distinct fracture sets are observed. Orientations of the three fracture sets in the Upper Spraberry as measured from 107 natural fractures from 300 ft of horizontal core. One NE/SW fracture set is observed in the upper 1U sand and 2 NNE and ENE fracture sets are observed in the lower 5U sand (from Baker et al. 2001).

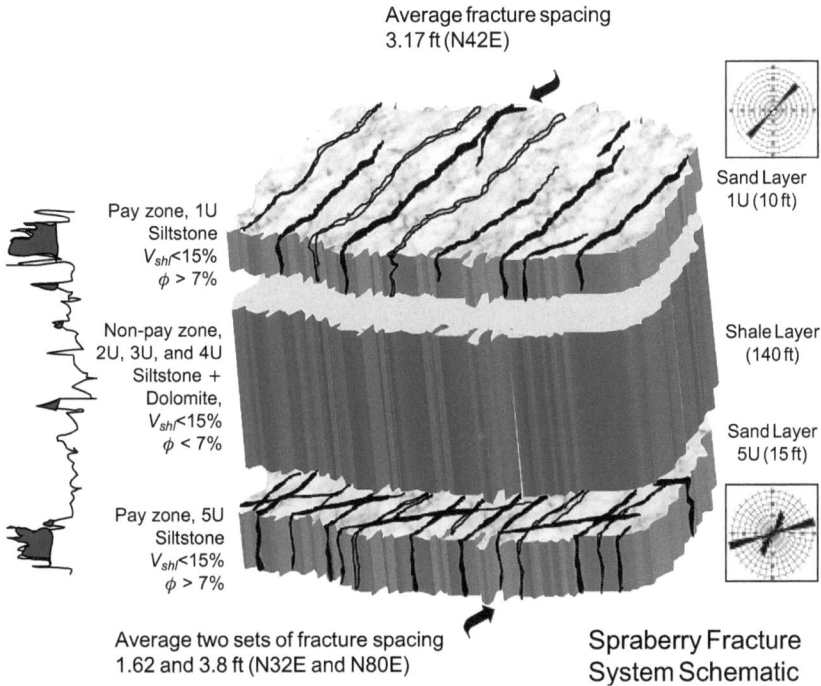

Average fracture spacing
3.17 ft (N42E)

Pay zone, 1U
Siltstone
V_{shl}<15%
ϕ > 7%

Non-pay zone,
2U, 3U, and 4U
Siltstone +
Dolomite,
V_{shl}<15%
ϕ < 7%

Pay zone, 5U
Siltstone
V_{shl}<15%
ϕ > 7%

Sand Layer
1U (10 ft)

Shale Layer
(140 ft)

Sand Layer
5U (15 ft)

Average two sets of fracture spacing
1.62 and 3.8 ft (N32E and N80E)

Spraberry Fracture
System Schematic

Fig. 5.7—Spraberry trend area fracture system. Three sets of fractures are shown in the upper 1U and lower 5U hydrocarbon saturated rock (after McDonald et al. 1997).

vertically within the reservoir units, and (2) hydraulic fractures are free to propagate at variable orientation that may not parallel the orientation of the natural fracture system (Schechter et al. 1996; Montgomery et al. 2000).

5.2.4 Multiwell Interference Testing. An interference test consisted of sequential injection of 2000 BWPD in Wells 47, 45, 25, and 48. Water was injected for 4 to 5 weeks into one well at a time, with a field stabilization period (2 to 3 weeks) between injection periods. Bottomhole pressure response was observed at Wells 38, 39, and 40 via permanent bottomhole pressure gauges in these wells. The pressure history is shown in **Fig. 5.8**.

The permeability ratios from reference points of Wells 47 and 45 obtained are 1.87 and 69.5 respectively, as is shown in **Fig. 5.9**. The orientation of maximum permeability from reference points of Wells 47 and 45 is N65°E and N37°E, respectively. Baker et al. (2000) analyzed responses at Wells 39 and 40 and found that the maximum permeability orientation was 15° offset from Wells 40–47 (toward Well 39). These tests, once again, reflect the heterogeneity of the three fracture sets as variable orientation, and anisotropy is evident even within localized areas.

5.2.5 Tracer Survey. Six different chemical tracers were injected in each of the six water-injection wells for a relatively long period (> 9 hours) to ensure that sampling at the producing wells would capture the tracer slug even in the case of very rapid breakthrough (< 1 day). Of the 29 wells that were sampled, 15 wells showed tracer breakthrough within

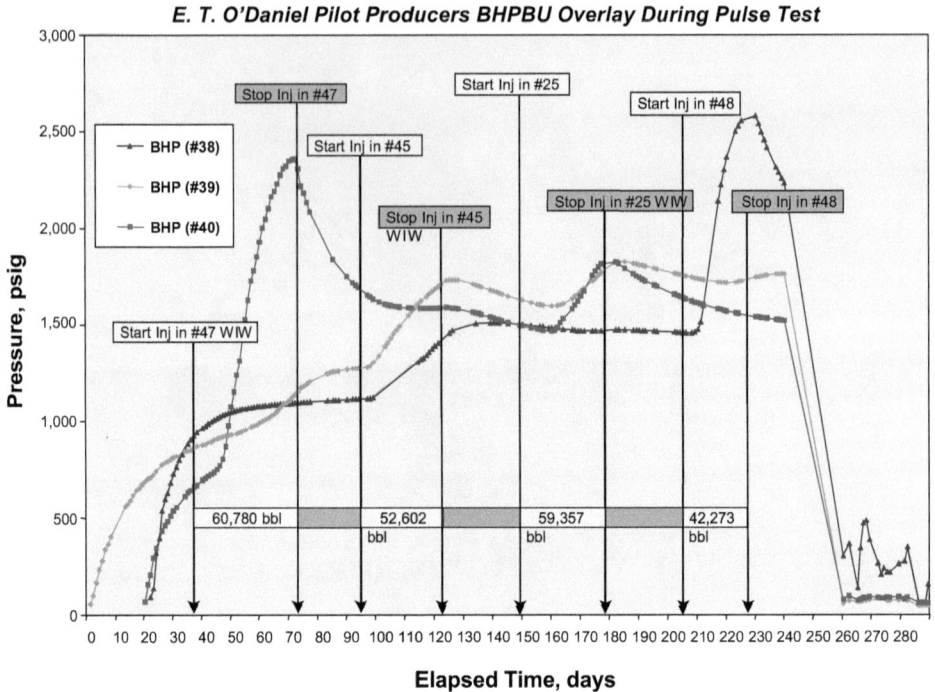

Fig. 5.8—Results of multiple well interference test from O'Daniel pilot area. Four pulses were initiated in Wells 47, 45, 25, and 48. Pressure was monitored with PBHP gauge assemblies in O'Daniel Wells 38, 39, and 40 (from Baker et al. 2000).

the first 2 days after initiation of tracer injection. This indicates that the fracture system has very high permeability that enables rapid fluid transport over significant distances (more than 3,000 ft in 2 days), reaching from Floyd "B1" to Brunson "D1." Three production wells returned concentrations above 150,000 ppt. The injector-producer well pairs are as follows:

1. WIW Well 47 to Brunson D-1 (N33°E, 3,200 ft)
2. WIW Well 46 to Brunson D-1 (N37°E, 4,480 ft)
3. WIW Well 48 to O'Daniel A-1 (N57°E, 1,880 ft)
4. WIW Well 45 to Pilot Well 38 (N74°W, 640 ft)

From three observation wells in the pilot area, only Pilot Well 38 showed significant tracer breakthrough, while slight tracer breakthrough is observed in Pilot Wells 39 and 40. This behavior is consistent with results previously reported that fluid flow occurs primarily along a NE/SW orientation corresponding to fractures seen in 1U unit cores. Local variation in fracture orientation and permeability anisotropy is also possible depending on the contribution from the two sets of fractures observed in 5U core. Fig. 5.5 summarizes the tracer and interference tests. Note that Injection Well 45 indicates anisotropic flow during the pressure interference test along the 1U fracture orientation, but high tracer concentration was observed in the production well oriented E/W from Well 45. In Well 47, the inter-

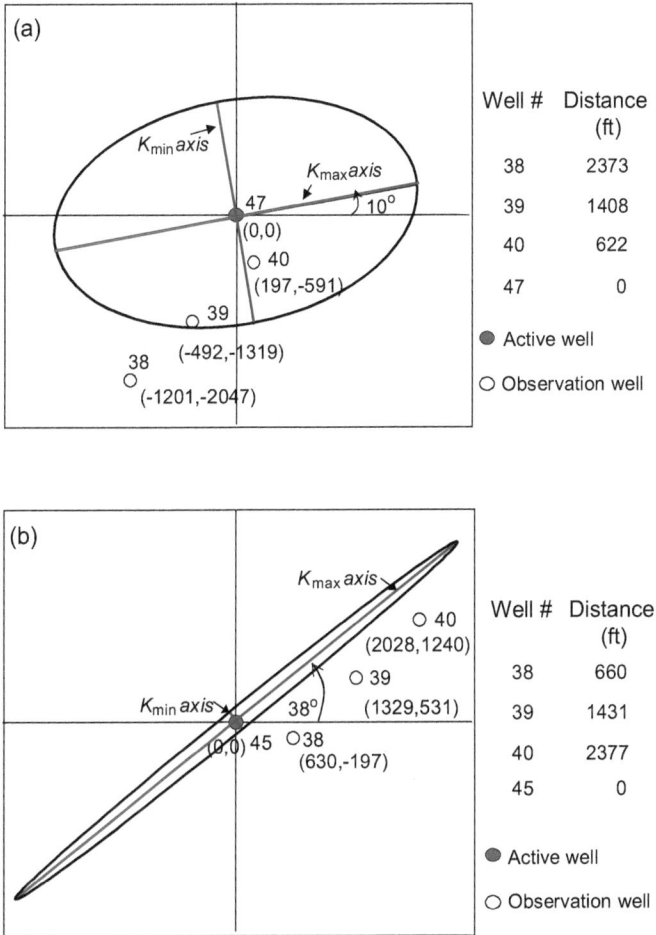

Fig. 5.9—(a) Degree and orientation of anisotropy around Injection Well 47. Note the moderate anisotropy and orientation consistent with the ENE set of fractures in the 5U. (b) Degree and orientation of anisotropy around Injection Well 45. Anisotropy is high, and orientation is consistent with the NE set of fractures in the 1U (after Baker et al. 2000).

ference test indicates low anisotropy; however, the tracer shows from this well are more in agreement with the fracture set seen in the 1U. This seemingly contradictory performance is explained by injection profile logging.

5.2.6 Injection Profiles. Several injection profiles were measured in the water- and gas-injection wells. Results indicate that flow into zones within the upper Spraberry is variable. For instance, an early profile log may indicate that 100% losses occur in the 1U zone, while a later test may indicate that flow distribution has changed and fluid loss occurs mainly through the 5U. Since each zone has different fracture characteristics, it can now be explained why tracer response and interference testing leads to variable anisotropy and the contradictory results when the interference and tracer tests are compared.

In summary, pre-water-injection tests have demonstrated the following characteristics of the Spraberry fracture system:

- The fracture system consists of three fracture sets in two layers, and, on the basis of both injection and interference tests, the response of neighboring wells differs depending on which layer receives the injected fluid.
- Permeability anisotropy, once considered the result of cross-fractures or matrix permeability, is probably the result of the multiple fracture sets and intersections of the multiple fracture sets. The variability in distribution of each fracture set relative to an injection well and the injection profile would then determine local permeability anisotropy.

5.2.7 Results of Water Injection. Stable water injection was initiated in October 1999 in order to increase the reservoir pressure above the minimum miscible pressure (MMP). Another goal was establishment of a baseline decline so that all produced oil as a result of CO_2 injection could be quantified. Since this was a previously waterflooded area, it was assumed that the reservoir was at residual oil saturation and that there would be minimal response from water injection. Production has been recorded in 37 wells in and around the pilot area. Oil production was divided into 23 off-trend and 7 on-trend wells. The seven on-trend wells are located along the NE/SW natural fracture orientation as determined by horizontal core analysis.

The three central production wells (38, 39, and 40) showed small incremental oil production after initiation of water injection. However, select wells oriented parallel to the primary fracture trend responded with a significant increase in oil production (**Fig. 5.10**). Some production wells over 1 mile from the injection wells have responded, whereas production wells located a fraction of that distance off the predominant fracture trend from injection wells have shown little response, similar to that observed in the tracer survey and previously reported waterflood results. The unexpected increase in oil production along the fracture trend indicates clearly the presence of unswept oil even though the area had been waterflooded in the 1960s.

The composite performance of the "expanded" pilot area shown in **Fig. 5.11** indicates that oil production has steadily increased from 200 BOPD before water injection to a current level near 400 BOPD with an additional cumulative production of approximately 150,000 bbl of oil after 3 years of water injection. Thus, the seven on-trend wells are responsible for 150 to 200 bbl of incremental oil per day. This represents an incremental gain of 20–30 barrels for each of seven on-trend wells. The results also dispel the notion that on-trend injection wells will channel via the fractures and rapidly water-out production wells without producing incremental oil by means of the imbibition mechanism.

In summary, Spraberry fractures cause flow anisotropy, and the matrix permeability is very low, thus requiring high-rate fluid injection to force fluids to a line of production wells from a parallel line of injection wells. The injection rate required to force fluids to off-trend wells is above the parting pressure of the rock and creates hydraulic fractures. The small difference in stress anisotropy between adjacent sand and shale layers results in paths or preferential channeling through the intervening nonpay shaly rock, resulting in poor sweep efficiency. Putra et al. (1999) discuss the low-rate injection required to (1) match the injection rate to the rate of imbibition and (2) keep below the parting pressure and confine the injected fluids to the reservoir rock. This low-pressure differential is unable to build enough

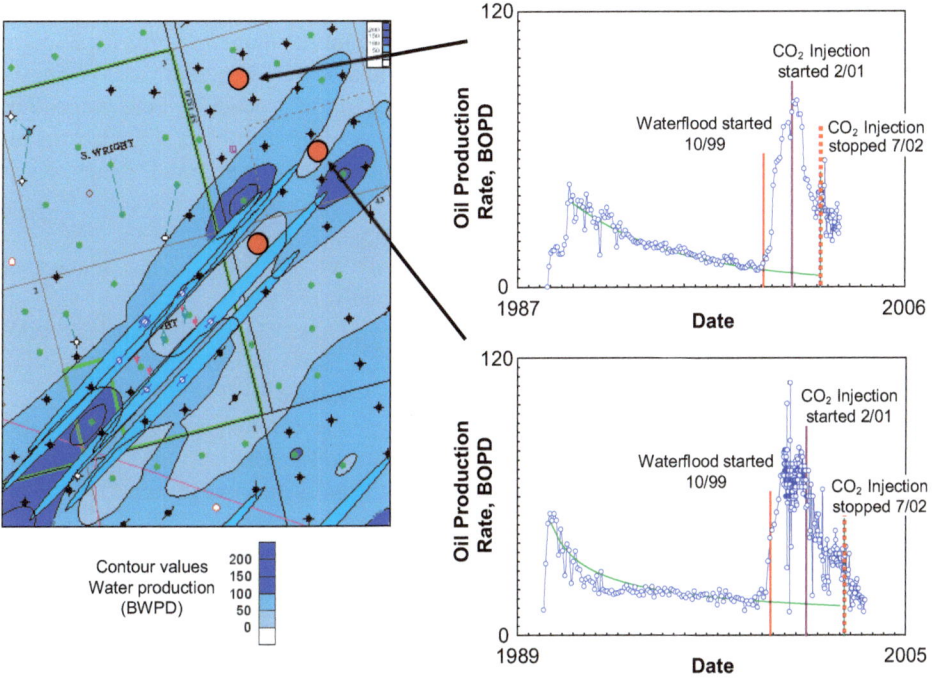

Fig. 5.10—Response to water injection in the Spraberry trend area in wells located 1 mile away from water injectors along the NE/SW fracture trend. Map contours show water cut; map ellipses show hypothetical distribution of injected water.

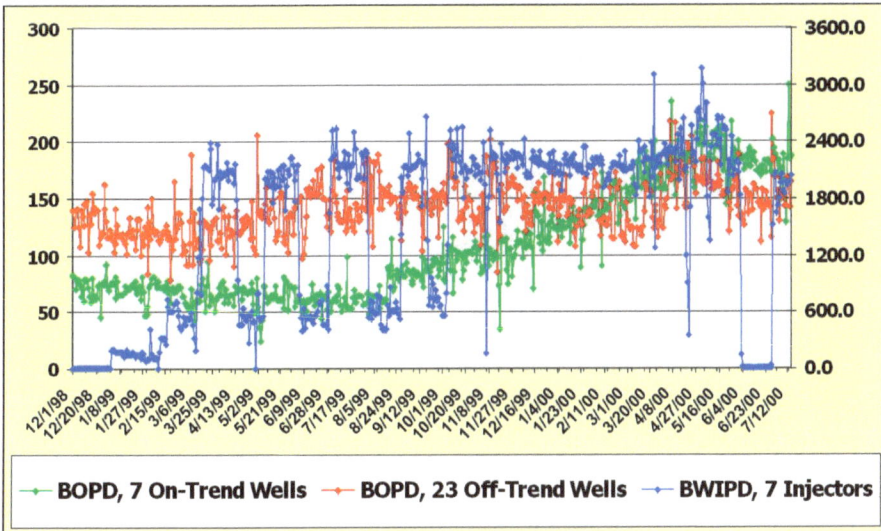

Fig. 5.11—Composite response of on-trend and off-trend injection wells to water injection in the Spraberry pilot area. Note the lack of response in wells perpendicular to the fracture trend and significant oil response in wells parallel to the fracture trend.

interference between injection wells to force fluid to off-trend wells as occurs at Midale field. The net conclusion is that fluid travels faster and farther along the primary fracture orientation than previously believed. On typical 40- and 80-acre injection well spacing in Spraberry, a line of injection wells might never affect a line of production wells perpendicular to the fracture trend. This fact resulted in operators being perplexed at the lack of response in the interior part of a pattern, while wells along the fracture trend, sometimes at great distances from injectors, would see a significant response in oil production.

5.2.8 Results of CO_2 Injection. Upon initiation of CO_2 injection, almost immediate breakthrough was observed in the central production wells. The CO_2 seems contained, with only two wells outside of the pilot area producing CO_2. Other wells in the expanded pilot area are monitored for CO_2, yet there is no indication of breakthrough. Incremental oil is observed in two offset wells (O'Daniel A Well 1 and O'Brien B Well 1).

Two observation wells cased in fiberglass were logged at different time intervals to understand distribution of CO_2 away from the injection wells. The results from O'Daniel Well 49 indicate that injected CO_2 was present throughout the gross interval. In contrast with expectations, this indicates that CO_2 was not contained vertically.

Difficulty in interpreting the CO_2 injection remains, since the original slug volume was designed based on a 20-acre confined pattern. The total design volume was 128 MMscf, or 80,000 res bbl. After injection of water was initiated, it was realized that the actual rock volume being swept was far greater than 20 acres. Approximately 130,000 bbl of water were required before a response was observed during water injection. The volume of injected CO_2 required before a significant hydrocarbon pore volume is attained may preclude CO_2 injection on economic grounds. Also, the small volume of incremental oil expected over such a large area, combined with the unexpected positive results from water injection, tends to obscure incremental oil recovery, making it difficult to assess utilization. The results of the pilot thus far should be sufficient to trigger more interest in pursuing injection strategies in the Spraberry trend area.

5.3 Ekofisk Field

Ekofisk is a giant oil field producing from a low-permeability naturally fractured chalk reservoir located in the Norwegian sector of the North Sea (**Fig. 5.12**). The field was discovered in 1969, started to produce in 1971, and began a fieldwide waterflood program in 1987. Production rates in 1998 were approximately 300,000 BOPD and 500,000 Mscf/D from 76 wells under waterflood (Agarwal et al. 2000).

Primary recovery is estimated to be 18% of OOIP and was considered successful despite initial concerns over permeability loss associated with pressure depletion and the potential for stress-induced closure of the natural fractures in the reservoir. Although the fracture system represents a very small portion of the hydrocarbon pore volume, it has significantly affected production by increasing effective permeability.

Enhanced recovery studies were initiated soon after the start of primary production to evaluate the fractured chalk behavior and its impact on expected recovery. Then a major field study commenced in 1991 to evaluate the operating strategy for the next recovery stage.

5.3.1 Geology. The main reservoirs of Ekofisk field are high-porosity limestone chalks with average matrix permeability of 1 md. The field is 12,000 acres in size, and it is located on a N/S-trending anticline (**Fig. 5.13**). It produces from two naturally fractured chalk units: the

Fig. 5.12—Ekofisk field location in North Sea (after Jensen et al. 2000).

Danian (early Paleocene) age Ekofisk formation and the Maastrichtian age (latest Creta-ceous) Tor formation. These are separated by a thin, impermeable clay-rich layer. The Ekofisk formation is 350 to 500 ft thick, and the Tor formation is 250 to 500 ft thick. Both of these formations have an average porosity of approximately 30%. Approximately two-thirds of the 6.4 billion STB hydrocarbon pore volume in place is in the Ekofisk formation.

5.3.2 Reservoir and Fluid Properties. The initial reservoir pressure of Ekofisk field was 7,135 psia at a depth of 10,400 ft. The original reservoir fluid was undersaturated volatile oil with a bubble point pressure of 5,560 psia at an initial temperature of 268°F. In 1976, the reservoir pressure decreased below the bubblepoint, so that by 1986, the average GOR had increased from an initial solution GOR of 1,530 scf/STB to 9,000 scf/STB. **Table 5.3** summarizes the Ekofisk field reservoir parameters.

5.3.3 Fracture Analysis. Fractures were studied in more than 3,000 ft of core from 14 wells throughout the field, some of which were oriented. Additionally, fractures were in-terpreted on resistivity imaging logs from 11 wells. Fractures were examined for their char-acteristics, orientation, spacing (density), and lithologic association. This information was used to understand fracture distribution (spatially and lithologically) and orientation in order to determine the orientation and relative magnitude of permeability anisotropy, and the geometry of the matrix blocks for dual-porosity simulation.

Three main fracture types have been associated with enhanced permeability at Ekofisk:

- Fractures associated with stylolites.
- Large-scale (seismically detectable) faults.
- Tectonic fractures.

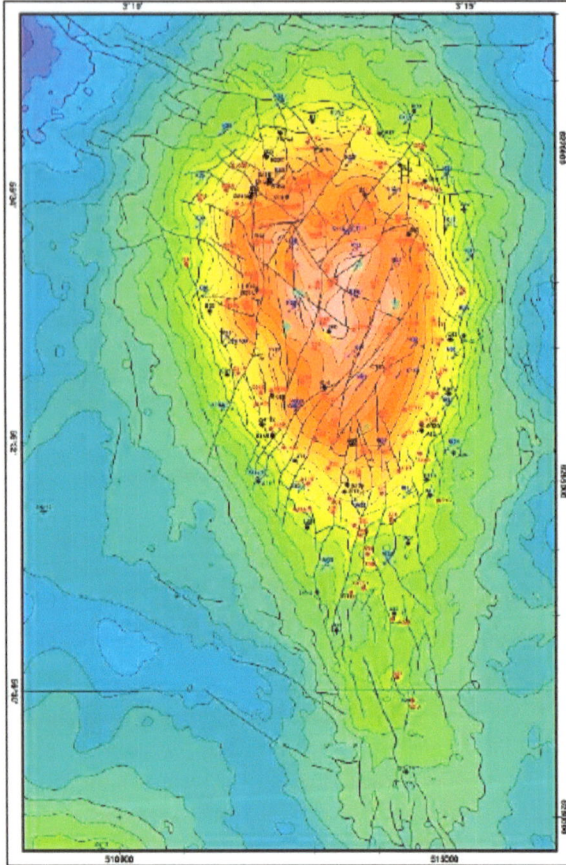

Fig. 5.13—Top structure map of Ekofisk field (after Jensen et al. 2000).

Each has distinct characteristics, modes of origin, occurrence, and impact on fluid flow. Each fracture type can be open or sealed by mineralization, generally calcite.

The stylolite-associated fractures are common in the Tor formation but relatively uncommon in the Ekofisk formation. These fractures are perpendicular to bedding-parallel stylolites and extend to less than 20 cm from the tips of stylolite teeth **(Fig. 5.14)**. They form a network of anatomizing fracture surfaces on both sides of the stylolites. They are well connected in the vicinity of stylolites and create bedding-parallel zones of enhanced permeability. However, their restricted occurrence is not significant to large-scale flow throughout the reservoir, and they are probably important mainly for their local effect on reservoir drainage (increased fracture surface area).

Northeast-trending normal faults are the dominant seismically detectable fractures in Ekofisk field. These faults may be responsible for localized regions, where pressure communication has been detected across the tight zone that separates the two reservoir formations.

Tectonic fractures are the main reservoir-permeability-enhancing fractures in the field. These fractures are planar, range in dip from near vertical to about 65°, and can have large

TABLE 5.3—EKOFISK RESERVOIR PARAMETERS	
Lithology	Chalk
Depth at reservoir midpoint, ft	10,400
Area, acres	12,071
Net pay with $\phi > 15\%$ and $S_w < 50\%$, ft	590
Average porosity, %	31.7
Matrix permeability average, md	1
Matrix permeability range, md	0.1–10
Maximum well test permeability, md	200
Initial reservoir pressure at 10,400 ft, psig	7,120
Reservoir temperature at 10,400 ft, °F	268
Temperature gradient through the the reservoir, degrees/100 ft	2.5
Water saturation, %	23.6
Initial GOR, scf/STB	1530
Initial FVF, RB/STB	1.78
Oil gravity, °API	36

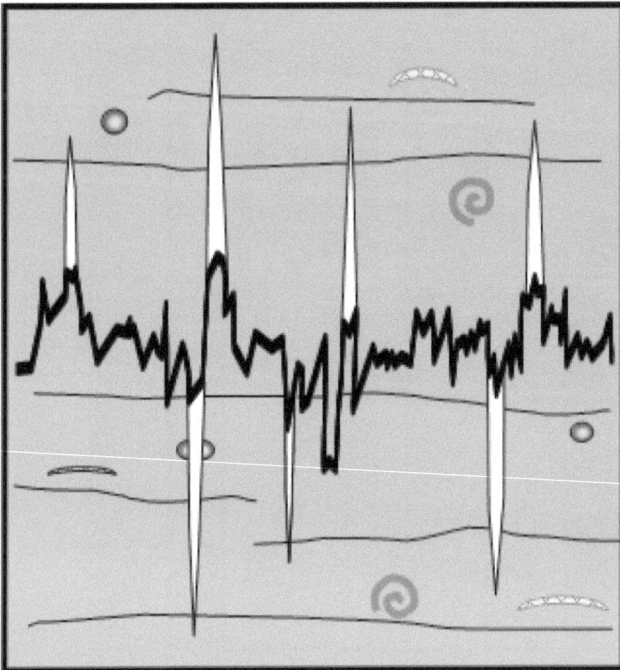

Fig. 5.14—Schematic cross section of stylolite-associated fractures with a vertical height that rarely exceeds 20 cm from stylolite surfaces. Dark, jagged curve represents a bedding-parallel stylolite surface; white peaks are fractures.

lengths (at least tens of meters) and heights on the order of 3 to 4 m (based on vertical well test results). Although fracture spacing (perpendicular distance between fractures) is highly variable, Dangerfield et al. (1992) state that it rarely exceeds 1 m anywhere in the field. Fritsen and Corrigan (1990) describe the tectonic fractures as small-scale normal faults. Dangerfield et al. (1992) mention the presence of both "fractures" and "faults," but assume that the tectonic fractures and seismic-scale faults are genetically related.

The tectonic fracture strike orientation is dominantly NE/SW, except in the east-central part of the field, where it is E/W (**Fig. 5.15**). Dangerfield et al. describe a secondary tectonic-fracture pattern that is "radial" with respect to the field, although this pattern is not strongly evident in Fig. 5.15. Some correspondence in strike is evident between the tectonic fractures and the seismic-scale faults (compare Fig. 5.13), but the correspondence breaks down on the east side of the field.

Fritsen and Corrigan (1990) used a set of wireline-log petrophysical characteristics to define eight distinct petrofacies in the Ekofisk and Tor reservoir rocks, and then looked for their correlation with fracture density. They found that fracture spacing ranged from a maximum of 33.6 cm in the least fractured log facies to 4.1 cm in the most densely fractured log facies (**Fig. 5.16**). They also correlated fracture height with the log facies. These parameters were then used as the basis for defining internal matrix block sizes as a basic input to a dual-porosity simulation model.

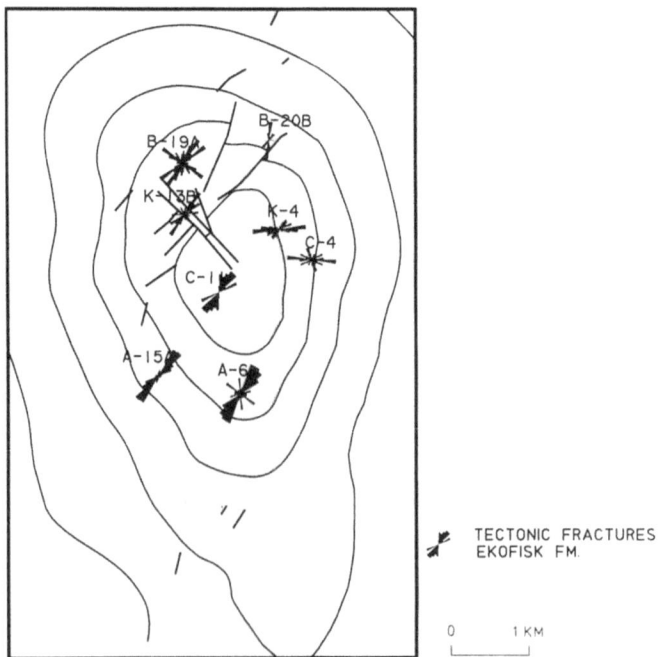

Fig. 5.15—Strike of tectonic fractures measured in oriented core from the Ekofisk formation (from Dangerfield et al. 1992). Reprinted from Dangerfield, J., Knight, I., and Farrell, H., Characterization of Faulting and Fracturing in Ekofisk Field From Seismic, Core and Log Data, in *Structural and Tectonic Modelling and Its Application to Petroleum Geology*, 397–407, © 1992, with permission from Elsevier.

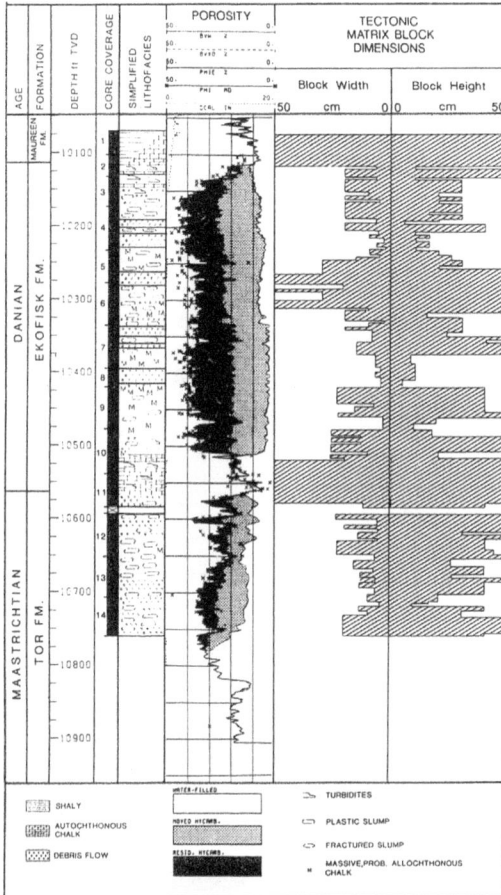

Fig. 5.16—Matrix block dimensions defined from fracture density in Ekofisk Well 2/4A-8 (from Fritsen and Corrigan 1990). Note that block width (= fracture spacing) is inversely proportional to fracture density.

5.3.4 Fracture Intensity, Porosity, and Position on the Structure. Agarwal and Allen (1996) found a relationship between fracture intensity for the stylolite-associated fractures and matrix porosity **(Fig. 5.17)**. For this fracture type, it is evident that fracture intensity on the flanks of the structure is greater than at the crest. On the other hand, Meling and Lehne (1993) show that the greatest fracture-induced flow enhancement in the field is near the crest of the structure, and that the flow enhancement decreases down-flank. They interpret this observation in terms of fracture density, concluding that the greatest fracture density is in the crestal position at Ekofisk field and decreases down-flank.

This apparent disagreement may be a function of differences in fracture type. The data in Fig. 5.17 refer to stylolite-associated fractures in the Tor formation only, whereas Meling and Lehne (1993) base their conclusions on well-test data, which would be most strongly influenced by the tectonic fractures that are more important for reservoir flow. It is interesting that these two distinct fracture data sets exhibit distinctly different spatial distributions, an observation that serves to underscore the fact that no general model is presently adequate for predicting fracture occurrence in a reservoir.

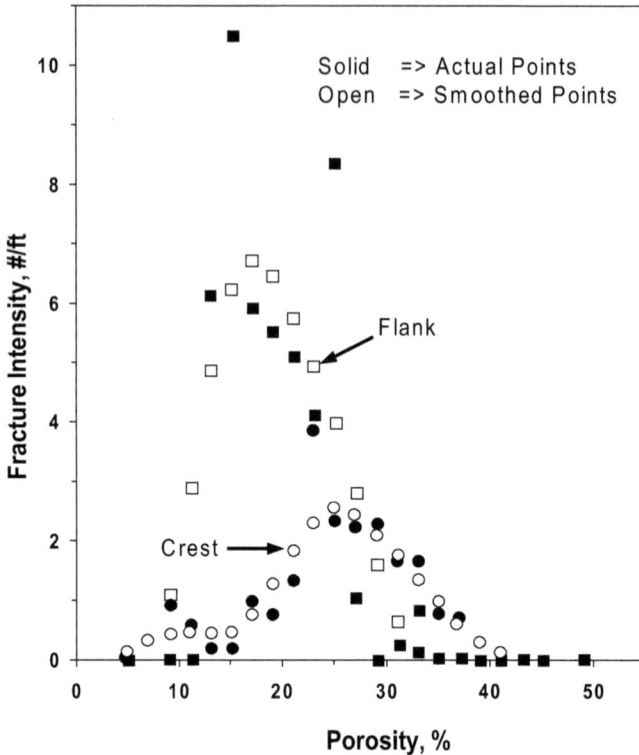

Fig. 5.17—Relationship between fracture intensity and porosity of stylolite-associated fractures in the Tor formation (from Agarwal and Allen 1996).

5.3.5 Well-Test Analysis. Meling and Lehne (1993) present a fieldwide fracture index derived by dividing permeability from well test by matrix permeability for the completed interval (FI = k_{wtest}/k_{core}, which we defined as FCI in Chapter 3). Their measurements show moderate fracture indices, ranging from 5 to 30. They do not present details of their analysis, but it is likely that the increased permeability is integrated over a thick completion interval; they probably make the common assumption that the height of the producing interval approaches the height of the reservoir.

5.3.6 Enhanced Oil Recovery. Enhanced oil recovery studies in Ekofisk were initiated soon after the start of primary production to evaluate the effect of the fractures on recovery. A 1979 laboratory study of waterflood potential indicated favorable water imbibition into the Tor formation. Favorable results from a water injection pilot that began in 1981 led to a waterflood of the northern Tor formation that began in 1983.

The positive results of Tor water injection encouraged a water-injection pilot into the lower third of the Ekofisk formation. A successful pilot project from 1985 to 1987 concluded that the lower Ekofisk formation responded as favorably as the Tor formation. Therefore, the Tor formation waterflood was expanded to a fieldwide project, and water injection into the Lower Ekofisk formation was undertaken as well.

5.3.7 Laboratory Data. The imbibition data used for the waterflood pilot study were obtained from laboratory experiments performed in the late 1970s. However, there is no general correlation between imbibition and porosity available for either the Upper or Lower Ekofisk. Although some samples indicated imbibition in excess of the Tor value, the majority of samples showed imbibition in the range of 15 to 25%. The Lower Ekofisk in the southern portion of the field showed higher imbibition values, averaging close to those of the Tor. Additionally, Ekofisk samples tested with large gas saturations indicated imbibition values approaching those of the Tor. The lowest imbibition values in the Lower Ekofisk were measured in the pilot area, which averages only slightly above 10% (**Fig. 5.18**).

5.3.8 Analysis of Core Data. Based on 50 sets of mercury capillary pressure measurements, the irreducible wetting phase saturations were estimated at around 10 and 6 percent for Ekofisk and Tor formations, respectively. Regression analysis of capillary pressure plot vs. reduced saturation provides estimates for the apparent pore throat distribution coefficient, λ, with an average value of 2.52 for Ekofisk and 2.20 for the Tor formation.

As the waterflood progressed, it became clear that the water retention in the reservoir was in excess of the laboratory-derived imbibition endpoint. The long period of water-free production at wells B-24 and B-19, as well as low water cut in well B-22, in combination with openhole logging results on well B-16, necessitated revision of the Ekofisk imbibition data.

An experiment was conducted to better define the negative portion of the imbibition capillary pressure curve. The result suggested that a significant volume of additional oil could be mobilized in the Ekofisk formation with small pressure differential across the core. A typical curve is displayed in **Fig. 5.19**.

5.3.9 Water/Oil Imbibition Data. Water/oil imbibition experiments were performed on Maastrichtian cores from several wells. Measurements were made on 1-in.-diameter cores

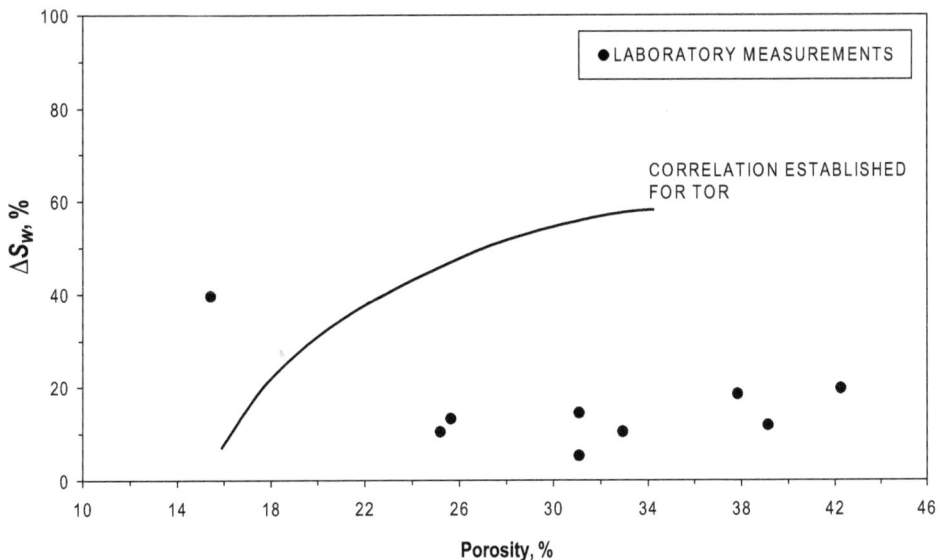

Fig. 5.18—Imbibition data of the Lower Ekofisk, Well B-16 (after Sylte et al. 1988).

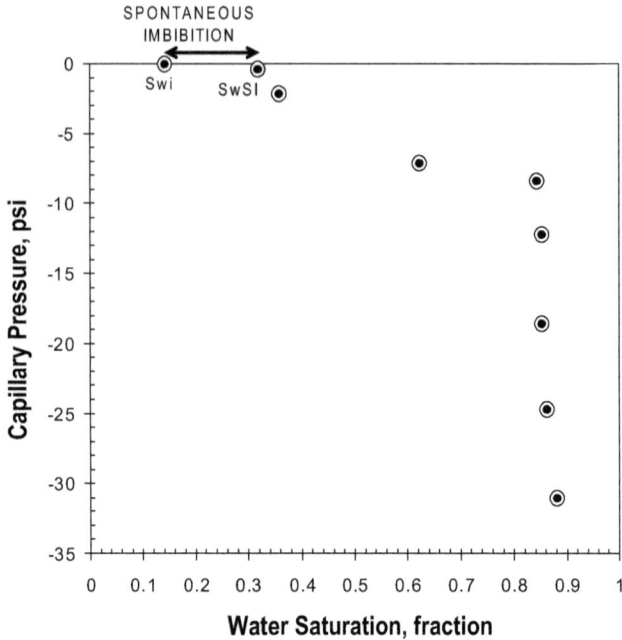

Fig. 5.19—Water saturation in an Ekofisk core as a function of negative capillary pressure (after Sylte et al. 1988).

that were either preserved or extracted and restored. The experiments were conducted in three stages: (1) measurements of imbibition data, (2) measurements of oil recovery vs. time, and (3) direct measurements of imbibition capillary pressure data. Typical experimental results for these procedures are shown in **Figs. 5.20 through 5.22**.

5.3.10 Variation in Free Water Level. Each well examined in the field shows an individual free water level (FWL). The differences between wells are significant and vary up to 200 m. The deepest contact was found on the flanks, and the shallowest contacts were at the crest. The FWL tilts in the direction perpendicular to the fault and fracture direction. There were two suggestions regarding this variation in FWL: (1) It was due to variations in rock types vs. depth, and (2) it was due to the wetting phase barriers created by major fracture zones.

5.3.11 Identification of Capillary Pressure Systems and Major Fracture Zones. The calculation of erratic capillary pressures in fractured intervals shows that zones with fewer fractures have linear capillary pressure vs. depth. Several capillary pressure systems were identified, which probably are separated by major fracture zones.

5.3.12 Tor Formation Waterflood Pilot. The first Ekofisk-field waterflood pilot is located in the northwest corner of the field **(Fig. 5.23)** and includes Wells B-08, B-14, B-16, B-18, B-19, and B-21 through B-24.

5.3.13 Well Tests. An interference test was performed in January 1981 using Well B-22 as the active well and wells B-16 and B-24 as the observation wells. The objective was to de-

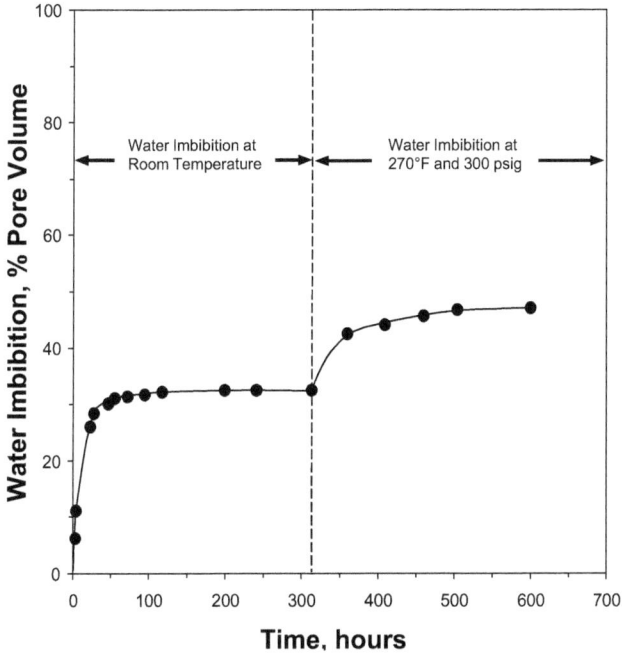

Fig. 5.20—Water/oil imbibition data (after Thomas et al. 1987).

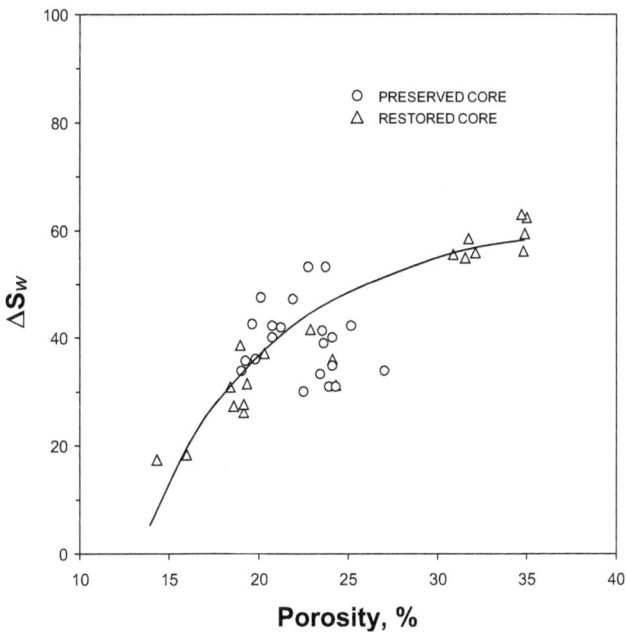

Fig. 5.21—Incremental water saturation vs. porosity (after Thomas et al. 1987).

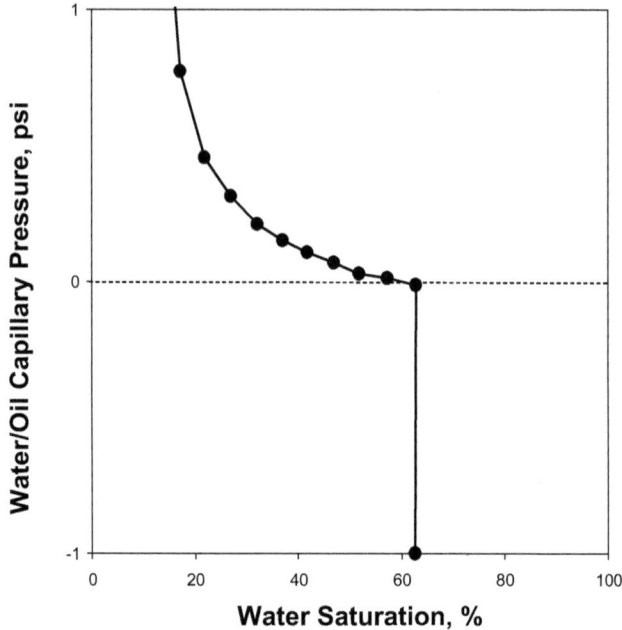

Fig. 5.22—Water/oil imbibition capillary pressure vs. water saturation (after Thomas et al. 1987).

termine k_{max} and k_{min}. The result of this test is summarized in **Table 5.4**. This result indicates a fairly strong anisotropy, with the maximum permeability direction oriented NE/SW, similar to the orientation of the tectonic fractures (Fig. 5.15).

5.3.14 Wettability. The wettability is proportional to the slope of the calculated capillary pressure. For the gas/condensate system, the $\sigma \cos \theta$ at reservoir conditions was estimated to be about 60 mN/m. For the oil field, the data show a weak capillary force with $\sigma \cos \theta$ on the order of 7 mN/m (Meling and Lehne 1993).

During this waterflood pilot study, the following conclusions were drawn: (1) Water imbibition into Maastrichtian chalk is high (up to 45% pore space in core with 25% porosity); (2) trapped gas following water imbibition will be less than 8%, and trapped gas following oil imbibition is less than 4% (this gas saturation is reduced to zero following water imbibition); (3) gas/oil capillary continuity exists across the fractures in the Tor formation; (4) analysis of core and well-test data indicates that fracture spacing and matrix block size in the fractured fraction of the Maastrichtian is 1 to 3 ft; and (5) interference tests indicate that a fast anisotropy direction oriented NE/SW exists in the pilot area. All of these values were used in the simulation of the Tor pilot, in which an excellent history match was achieved.

5.3.15 Ekofisk Formation Waterflood Pilot. The Ekofisk formation was initially given lower waterflood priority based on its less-favorable response in spontaneous imbibition laboratory experiments. However, the large value of initial oil in place—in combination with the recognition that mechanisms other than imbibition may play a role in waterflood performance—led the operators to perform a pilot in the Lower Ekofisk. The pilot was also designed to evaluate rock stability and injectivity in this fractured chalk.

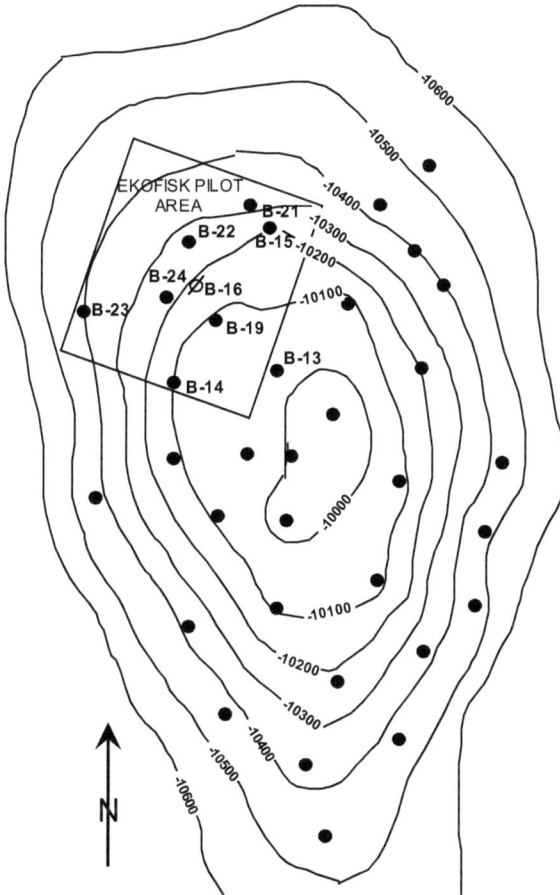

Fig. 5.23—Structure of top of Maastrichtian and the location of Tor waterflood pilot (after Thomas et al. 1987).

TABLE 5.4—RESULT OF INTERFERENCE TEST IN JANUARY 1981 (after Thomas et al. 1987)	
Parameter	Value
k_{max}, md	220
k_{min}, md	32
Azimuth, degrees	022

The pilot consisted of one injector (Well B-16) and three producers (Wells B-19, B-22, and B-24) in an unconfined four-spot pattern (**Fig. 5.24**).

5.3.16 Interference Tests. Interference testing in the Ekofisk formation pilot was designed to determine anisotropy based on the regional permeability variations and local fracture trends between the pilot wells. The interference test utilized the central Well B-16 as the active well and Wells B-19, B-22, and B-24 as the observation wells. Pressure responses were

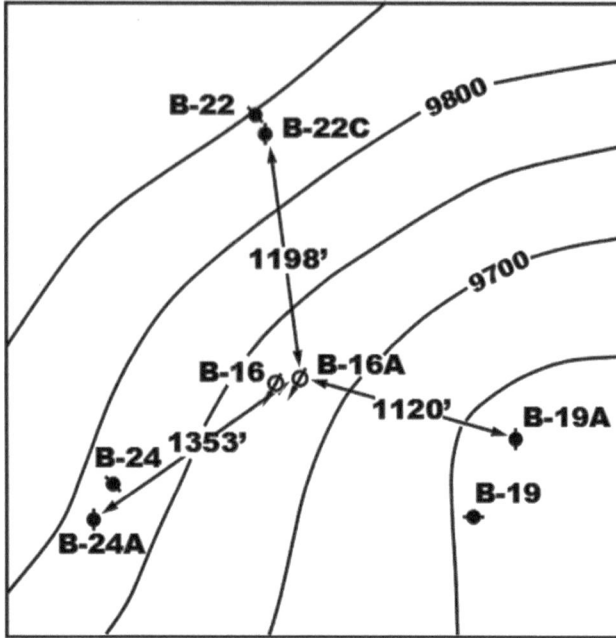

Fig. 5.24—Lower Ekofisk pilot area (from Sulak et al. 1990).

recorded in the observation wells as B-16 was produced for 121 hours, then shut in for 168 hours. A static period of 72 hours was monitored prior to the activation of the active well.

Results indicated a rapid response of 12 hours between Wells B-16 and B-22, a slower response of 32 hours between Wells B-16 and B-19, and no response between Wells B-16 and B-24. A qualitative analysis indicated strong north-south anisotropy. This north-south anisotropy is underscored by the fact that well-test permeability is much less in the B-22 area than in the B-19 area. This is surprising, given that the strike of tectonic fractures is predominantly NE/SW in this portion of the field (Fig. 5.16), and it indicates the importance of in-situ dynamic tests of the reservoir-scale permeability field. More tests would be needed to determine whether this result is actually typical of the reservoir, or whether the results are a manifestation of fracture-related heterogeneities.

Interwell permeability was calculated using a line-source solution technique. Then the anisotropy analysis was done based on the assumption that k_{max} was in the direction of B-16 to B-22 (north-south). Results are tabulated in **Table 5.5**.

TABLE 5.5—RESULTS OF INTERFERENCE TEST ANALYSIS OF EKOFISK FORMATION PILOT (after Sylte et al. 1988)	
Parameter	Value
k_{max}, md	153
k_{min}, md	82.4
Azimuth, degrees	108

	kh/μ		Pressure
Well	(md-ft/cp)	Skin	(psig)
B-19	9,936	–4.2	4,025
B-22	2,117	–4.3	3,995
B-24	4,370	–4.0	3,850

TABLE 5.6—WELL-TEST RESULTS AFTER INITIAL COMPLETIONS (after Sylte et al. 1988)

5.3.17 Well Tests. Extensive pressure testing and production logging has been performed on each Ekofisk pilot producer. The results demonstrate a large permeability variation within the pilot area, reinforcing the impression of reservoir heterogeneity indicated by the interference-test results. The skin values obtained indicate that well stimulations were effective **(Table 5.6)**.

5.3.18 Radioactive Tracer Observation. The first and primary radioactive tracer used in the Ekofisk pilot was tritium, which was injected continuously from 18 August 1986, with the injected seawater at a constant concentration. A second tracer used was iodine-125, which was injected as a slug midway through the injection period.

Tritium production was first observed in Well B-22 after 50 days of first-tracer injection. The response was then observed in Well B-19 after 170 days of injection. Iodine-125 also was first observed in Well B-22 21 days after the start of iodine-125 injection. There is no response from the tritium and iodine-125 observed in Well B-24, which confirms earlier analysis that this well is not in direct communication with the injector.

5.3.19 Vertical Permeability. The vertical permeability of the pilot area was estimated to be 0.02 to 0.1 times the horizontal permeability, based on well performance and core analysis. The high initial GOR in wells perforated at the bottom of the Ekofisk indicates low vertical permeability. Gas saturations and associated GOR are somewhat higher in the upper layer than in the lower layers.

Although core analysis indicates that the majority of fractures are subvertical, in general they are believed to be limited in height to 4 m or less. The low in-situ vertical permeability concurs with the observation that fracture height is restricted.

5.3.20 Conclusions Regarding Enhanced Recovery. Results of the Ekofisk formation pilot indicate that the pilot area exhibits favorable waterflood characteristics and that residual oil saturation less than 40% can be expected.

Nomenclature

A_f	=	area of an individual fracture
b	=	bulk
C	=	system compressibility, psi^{-1}
d_f	=	fracture density
f	=	fracture
g	=	gravitational constant, m/s^2
h	=	height of producing interval
h_f	=	vertical height of fracture
H_{av}	=	stratigraphic thickness
k	=	permeability, md
K_m	=	matrix permeability
L	=	characteristic length of fracture spacing distribution, ft
m	=	matrix
n_f	=	total number of fractures sampled
N	=	number of parallel set of fractures (1, 2, or 3)
p	=	fluid pressure
p_f	=	fluid pressure in fracture
p_m	=	fluid pressure in matrix
q	=	transfer function
r_w	=	wellbore radius, ft
S_f	=	fracture spacing
t_D	=	dimensionless time
V_r	=	total volume of reservoir, ft^3
θ	=	contact angle
λ	=	interporosity flow coefficient, dimensionless
μ	=	fluid viscosity
μ_g	=	geometric mean of water and oil viscosities
μ_w	=	water phase viscosity
ρ	=	density, lbm/ft^3
σ	=	shape factor, 1/L^2
ϕ	=	porosity
ϕ_f	=	fracture porosity
ω	=	storativity ratio, dimensionless

References

Agarwal, B. and Allen, L.R. 1996. Ekofisk Field Reservoir Characterization: Mapping Permeability Through Facies and Fracture Intensity. Paper SPE 35527 presented at the European 3-D Reservoir Modeling Conference, Stavanger, 16–17 April.

Agarwal, B., Hermansen, H., and Sylte, J.E. 2000. Reservoir Characterization of Ekofisk Field: A Giant, Fractured Chalk Reservoir in the Norwegian North Sea—History Match. *SPEREE* **3** (6): 534–543. SPE-68096-PA.

Antonellini, M. and Aydin, A. 1994. Effect of Faulting on Fluid Flow in Porous Sandstones: Petrophysical Properties. *AAPG Bull.* **78:** 355–377.

Areshev, E.G., Dong, L.E., San, N.T., and Shnip, O.A. 1992. Reservoirs in fractured basement on the continental shelf of southern Vietnam. *J. Pet. Geol.* **15:** 451–464.

Bai, T. and Pollard, D.D. 2000. Fracture spacing in layered rock: A new explanation based on the stress transition. *J. Struct. Geol.* **22:** 43–57.

Baker, R.O., Bora, R., Schechter, D.S. et al. 2001. Development of a Fracture Model for Spraberry Field, Texas, USA. Paper SPE 71635 presented at the SPE Annual Technical Conference and Exhibition, New Orleans, 30 September–3 October.

Baker, R.O., Contreras, R.A., and Sztukowski, D. 2000. Characterization of the Dynamic Fracture Transport Properties in a Naturally Fractured Reservoir. Paper SPE 59690 presented at the SPE Permian Basin Oil and Gas Recovery Conference, Midland, Texas, 21–23 March.

Barfield, E.C., Jordan, J.K., and Moore, W.D. 1959. An Analysis of Large-Scale Flooding in the Fractured Spraberry Trend Area Reservoir. *JPT,* April, 15–19. SPE-1135-PA.

Barton, C.A., Zoback, M.D., and Moos, D. 1995. Fluid flow along potentially active faults in crystalline rock. *Geology* **23:** 683–686.

Beliveau, D. 1995. Heterogeneity, Geostatistics, Horizontal Wells, and Blackjack Poker. *JPT,* December, 1068–1074. SPE-30745-PA.

Beliveau, D., Payne, D.A., and Mundry, M. 1993. Waterflood and CO_2 Flood of the Fractured Midale Field. *JPT,* September, 881–887. SPE-22946-PA.

Bell, J.S. 1990. Investigating stress regimes in sedimentary basins using information from oil industry wireline logs and drilling records. *Geol. Soc. Spec. Publ.* **48:** 305–325.

Beur, K. and Trice, R. 2004. Data integration and numerical fracture models for water flood management: A case study of the Valhall Chalk Reservoir, North Sea. In R. Jolly, L. Lonergan, K. Rawnsley, and D. Sanderson (eds.), *Fractured Reservoirs*, Geol. Soc. London Spec. Publ. (in press).

Bourbiaux, B.J. and Kalaydjian, F.J. 1990. Experimental Study of Cocurrent and Countercurrent Flows in Natural Porous Media. *SPERE* **5** (3): 361–368. SPE-18283-PA.

Brownscombe, E.R. and Dyes, A.B. 1952a. Water-imbibition displacement: A possibility for the Spraberry. *Drill. and Prod. Prac.,* API, 383–390.

Brownscombe, E.R. and Dyes, A.B. 1952b. Water-imbibition displacement: Can it release reluctant Spraberry oil? *OGC,* 17 November, 264–265 and 377–378.

Cacas, M.C., Daniel, J.M., and Letouzey, J. 2001. Nested geological modelling of naturally fractured reservoirs. *Pet. Geosci.* **7:** 543–552.

Christie, R.S. and Blackwood, J.C. 1952. Production performance in Spraberry," *OGC,* 7 April, 107–115.

Craig, D.H. 1988. Caves and other features of Permian karst in San Andres dolomite, Yates Field Reservoir, West Texas. In N.P. James and P.W. Choquette (eds.), *Paleokarst,* 342–363. New York City: Springer-Verlag.

Dangerfield, J., Knight, I., and Farrell, H. 1992. Characterization of faulting and fracturing in Ekofisk Field from seismic, core and log data. In R.M. Larsen, H. Brekke, B.T. Larsen, and E. Talleraas (eds.), *Structural and Tectonic Modelling and its Application to Petroleum Geology,* Norwegian Pet. Soc. Spec. Publ. 1, 397–407. Amsterdam: Elsevier.

Deeg, W.E.J. 1998. Hydraulic Fracture-Initiation in Deviated or Horizontal Openhole Wellbores. Paper SPE 47386 presented at the SPE/ISRM Rock Mechanics in Petroleum Engineering, Trondheim, Norway, 8–10 July.

Dershowitz, B., LaPointe, P., Eiben, T., and Wei, L. 2000. Integration of Discrete Feature Network Methods With Conventional Simulator Approaches. *SPEREE* **3** (2): 165–170. SPE-62498-PA.

Dyes, A.B. and Johnston, O.C. 1953. Spraberry Permeability From Build-Up Curve Analyses. *Trans.,* AIME, **198:** 135–138.

Dyke, C.G., Wu, B., and Milton-Tayler, D. 1992. Advances in Characterizing Natural-Fracture Permeability From Mud-Log Data. *SPEFE* **10** (3): 160–166. SPE-25022-PA.

Elkins, L.F. 1953. Reservoir Performance and Well Spacing, Spraberry Trend Area Field of West Texas. *Trans.,* AIME, **198:** 177–195.

Elkins, L.F. and Skov, A.M. 1960. Determination of Fracture Orientation From Pressure Interference. *Trans.,* AIME, **219:** 301–304.

Elsayed, S.A., Baker, R., Churcher, P.L., and Edmunds, A.C. 1993. Multidisciplinary Reservoir Characterization and Simulation Study of the Weyburn Unit. *JPT,* October, 930–974.

Engelder, T. 1987. Loading paths to joint propagation during a tectonic cycle: an example from the Appalachian Plateau, U.S.A. *J. Struct. Geol.* **7:** 450–476.

Engelder, T., Gross, M.R., and Pinkerton, P. 1997. An analysis of joint development in thick sandstone beds of the Elk Basin Anticline, Montana–Wyoming. In T.E. Hoak, A.L. Klawitter, and P.K. Blomquist (eds.), *Fractured Reservoirs: Characterization and Modeling, RMAG Guidebook,* 1–18.

Enright, R.J. 1954. Imbibition: Newest producing technique. *OGC,* 17 May, 104–107.

Fritsen, A. and Corrigan, T. 1990.Establishment of a geological fracture model for dual porosity simulations on the Ekofisk Field. In Graham and Trotman (eds.), *North Sea Oil and Gas Reservoirs II,* Norwegian Inst. of Tech., 173–184.

Gale, J. 2002. Specifying Lengths of Horizontal Wells in Fractured Reservoirs. *SPEREE* **5** (3): 266–272. SPE-78600-PA.

Gauthier, B.D., Garcia, M., and Daniel, J.-M. 2002. Integrated Fractured Reservoir Characterization: A Case Study in a North Africa Field. *SPEREE* **5** (4): 284–294. SPE-79105-PA.

Gillespie, P.A., Howard, C.B., Walsh, J.J. and Watterson, J. 1993. Measurement and characterisation of spatial distributions of fractures. *Tectonophysics* **226:** 113–141.

Gilman, J.R. 2003. Practical aspects of simulation of fractured reservoirs. Keynote address, Intl. Forum on Reservoir Simulation, Baden-Baden, Germany, June.

Gross, M.R. 1995.Fracture partitioning: failure mode as a function of lithology in the Monterey Formation of coastal California. *Geol. Soc. Amer. Bull.* **107:** 779–792.

Guidroz, G.M. 1967. E.T. O'Daniel Project—A Successful Spraberry Flood. *JPT,* September, 1137–1140.

Haas, M.R., McFall, K.S., and Coates, J.M. 1989. Technology Advances Unlock the Potential of the Tight Frontier Formation in the Moxa Arch. Paper SPE 18962 presented at the SPE Rocky Mountain Regional/Low Permeability Reservoirs Symposium, Denver, 6–8 March.

Hailwood, E.A. and Ding, F. 1995. Palaeomagnetic reorientation of cores and the magnetic fabric of hydrocarbon reservoir sands. In P. Turner and A. Turner (eds.), *Paleomagnetic Applications in Hydrocarbon Exploration and Production,* Geol. Soc. London Spec. Publ. 98, 245–257.

Haller, D. and Porturas, F. 1998. How to characterize fractures in reservoirs using borehole and core images: case studies. In P.K. Harvey and M.A. Lovell (eds.), *Core-Log Integration*, Geol. Soc. London Spec. Publ. 136, 249–259.

Hanks, C.L., Wallace, W.K., Atkinson, P.K. *et al.* 2004. Character , relative age and implications of fractures and other mesoscopic structures associated with detachment folds: An example from the Lisburne Group of the northeastern Brooks Range, Alaska. *Bull. Cdn. Pet. Geol.* **52:** 121–138.

Heffer, K.J. and Lean, J.C. 1991. Earth stress orientation: A control on, and guide to flooding directionality in a majority of reservoirs. In B. Linville (ed.), *Intl. Reservoir Characterization Technical Conference III*, Tulsa, 800–822.

Hennings P., Olson J., and Thompson L. 2000. Combined outcrop data and structural models to characterize fractured reservoirs: An example from Wyoming. *AAPG Bull.* **84:** 830–849.

Hill, C.A. 1999. Origin of caves in the Capitan. In Saller, A.H., Harris, P.M., Kirkland, B.L., and Mazzullo, S.J. (eds), *Concepts in Sedimentology and Paleotology: Geologic Framework of the Capitan Reef*, SEPM Spec. Publ. 65, 211–222.

Hornby, B.E., Johnson, D.L., Winkler, K.W., and Plumb, R.A. 1989. Fracture evaluation using reflected Stoneley-wave arrivals. *Geophysics* **54:** 1274–1288.

Horne, R.N. 1995. *Modern Well Test Analysis: A Computer-Aided Approach,* second edition. Palo Alto, California: PetroWay.

Jaeger J.C. and Cook, N.G. 1979. *Fundamentals of Rock Mechanics,* third edition. London: Chapman and Hall.

James, N.P. and Choquette, P.W. (eds.) 1988. *Paleokarst.* New York City: Springer-Verlag.

Jensen, T.B., Harpole, K.J., and Østhus, A. 2000. EOR Screening for Ekofisk. Paper SPE 65124 presented at the SPE European Petroleum Conference, Paris, 24–25 October.

Kamal, M.M. 1983. Interference and Pulse Testing—A Review. *JPT,* December, 2257–2270.

Kulander, B.R., Dean, S.L., and Ward, B.J. 1990. Fractured core analysis: Interpretation, logging, and use of natural and induced fractures in core. *AAPG Methods in Exploration* **8.**

Lackie, M.A. and Schmidt, P.W. 1993. Drill core orientation using paleomagnetism. *Exploration Geophysics* **24:** 609–614.

Lee, S., Jensen, C., and Lough, M. 2000. An Efficient Finite Difference Model for Flow in a Reservoir With Multiple Length-Scale Fractures. *SPEJ* **5** (3): 268–275. SPE-65095-PA.

Lisle, R.J. 1994.Detection of zones of abnormal strains in structures using Gaussian curvature analysis. *AAPG Bull.* **78:** 1811–1819.

Lorenz, J.C., Sterling, J.L., Schechter, D.S., Whigham, C.L., and Jensen, J.L. 2002. Natural fractures in the Spraberry Formation, Midland Basin, TX: The effects of mechanical stratigraphy on fracture variability and reservoir behavior. *AAPG Bull.* **92:** 999–1030.

Luthi, S.M. and Souhaité, P. 1990. Fracture apertures from electrical borehole scans. *Geophysics* **55:** 821–833.

Ma, S., Zhang, X., and Morrow, N.R. 1995. Influence of fluid viscosity on mass transfer between rock matrix and fractures. Paper CIM 95-94 presented at the 46th Annual Technical Meeting of the Petroleum Soc. of CIM, Banff, Alberta, Canada, 14–17 May.

Marrett, R. 1997. Permeability, porosity, and shear wave anisotropy from scaling of open fracture populations. In T.E. Hoak, A.L. Klawitter, and P.K. Blomquist (eds.), *Fractured Reservoirs: Characterization and Modeling, RMAG Guidebook*, 217–226.

Mattax, C.C. and Kyte, J.R. 1962. Imbibition Oil Recovery From Fractured, Water Drive Reservoir. *SPEJ* **2** (2): 177–184. SPE-187-PA.

McDonald, P., Lorenz, J.C., Sizemore, C., Schechter, D.S., and Sheffield, T. 1997. Fracture Characterization Based on Oriented Horizontal Core From the Spraberry Trend Reservoir: A Case Study. Paper SPE 38664 presented at the SPE Annual Technical Conference and Exhibition, San Antonio, Texas, 5–8 October.

Meling, L.M. and Lehne, K.A. 1993.Description and Interpretation of north Sea Fractured Chalk Formation. Paper SPE 25640 presented at the 1993 SPE Middle East Oil Technical Conference and Exhibition, Bahrain, 3–6 April.

Montgomery, S.L. and Morgan, C.K. 1998. Bluebell Field, Uinta Basin: Reservoir characterization for improved well completion and oil recovery. *AAPG Bull.* **82:** 1113–1132.

Montgomery, S.L., Schechter, D.S., and Lorenz, J.C. 2000. Advanced reservoir characterization to evaluate carbon dioxide flooding, Spraberry Trend, Midland Basin, Texas. *AAPG Bull.* **84:** 1247–1273.

Mylorie, J.E. and Carew, J.L. 1995. Karst development on carbonate islands. In Budd, D.A., Saller, A.H., and Harris, P.A. (eds.), *Unconformities in Carbonate Strata—Their Recognition and the Significance of Associated Porosity*, AAPG Memoir **63:** 55–76.

Narr, W. 1991. Fracture density in the deep subsurface: techniques with application to Point Arguello oil field. *AAPG Bull.* **75:** 1300–1323.

Narr, W. 1996. Estimating average fracture spacing in subsurface rock. *AAPG Bull.* **80:** 1565–1586.

Narr, W. and Currie, J.B. 1982. Origin of fracture porosity: Example from Altamont field, Utah. *AAPG Bull.* **66:** 1231–1247.

Narr, W. and Suppe, J. 1991. Joint spacing in sedimentary rocks. *J. Struct. Geol.* **13:** 1037–1048.

National Park Service. 2004. *Cave Map of Wind Cave.*

Nelson, R.A. 2001. *Geological Analysis of Naturally Fractured Reservoirs*, second edition. Houston: Gulf Publishing Co.

Nelson, R.A., Lenox, L.C., and Ward, B.J. 1987. Oriented core: Its use, error, and uncertainty. *AAPG Bull.* **71:** 357–367.

Nelson, R.A., Moldovanyi, E.P., Matcek, C.C., Azpiritxaga, I., and Bueno, E. 2000. Production characteristics of the fractured reservoirs of the La Paz field, Maracaibo Basin, Venezuela. *AAPG Bull.* **84:** 1791–1809.

Nickelsen, R.P. and Hough, V.D. 1967. Jointing in the Appalachian Plateau of Pennsylvania. *Geol. Soc. Amer. Bull.* **78:** 609–629.

Oda, M. 1985. Permeability tensor for discontinuous rock masses. *Geotechnique* **35** (4): 483–495.

Ouenes, A., Richardson, S., and Weiss, W.W. 1995. Fractured Reservoir Characterization and Performance Forecasting Using Geomechanics and Artificial Intelligence. Paper SPE 30572 presented at the SPE Annual Technical Conference, Dallas, 22–25 October.

Putra, E., Fidra, Y., and Schechter, D.S. 1999. Use of Experimental and Simulation Results for Estimating Critical and Optimum Water Injection Rates in Naturally Fractured Reservoirs. Paper SPE 56431 presented at the SPE Annual Technical Conference and Exhibition, Houston, 3–6 October.

Ragan, D.M. 1985. *Structural Geology: An Introduction to Geometrical Techniques*, third edition. New York City: Wiley.

Reiss, L.H. 1980. *The Reservoir Engineering Aspects of Fractured Formations*. Houston: Gulf Publishing Co.

Rives, T., Razack, M., Petit, J.-P., and Rawnsley, K.D. 1992. Joint spacing: analogue and numerical simulations. *J. Struct. Geol.* **14:** 925–937.

Schechter, D.S., McDonald, P., Sheffield, T., and Baker, B. 1996. Reservoir Characterization and CO_2 Pilot Design in the Naturally Fractured Spraberry Trend Area. Paper SPE 35469 presented at the SPE Permian Basin Oil and Gas Recovery Conference, Midland, Texas, 27–29 March.

Schechter, D.S., Zhou, D., and Orr, F.M. Jr. 1994. Low IFT drainage and imbibition. *J. Pet. Sci. Eng.* **11:** 283–300.

Scholz, C.H., Dawers, N.H., Yu, J.-Z., and Anders, M.H. 1993. Fault Growth and Fault Scaling Laws: Preliminary Results. *J. Geophys. Res.* **98** (B12): 21,951–21,961.

Sulak, R.M., Nossa, G.R., and Thompson, D.A. 1990. Ekofisk field enhanced recovery. In Graham and Trotman (eds.), *North Sea Oil and Gas Reservoirs II*, Norwegian Inst. of Tech., 281–295.

Suppe, J. 1985. *Principles of Structural Geology*. Englewood Cliffs, New Jersey: Prentice-Hall.

Sylte, J.E., Hallenbeck, L.D., and Thomas, L.K. 1988. Ekofisk Formation Pilot Waterflood. Paper SPE 18276 presented at the SPE Annual Technical Conference and Exhibition, Houston, 2–5 October.

Thomas, L.K., Dixon, T.N., Evans, C.E., and Vienot, M.E. 1987. Ekofisk Waterflood Pilot. *JPT,* February, 221–232. SPE-13120-PA.

Thompson, L.B. 2000. Atlas of borehole imagery. *AAPG/Datapages* CD-ROM.

Verga, F.M., Carugo, C., Chelini, V., Maglione, R., and De Bacco, G. 2000. Detection and Characterization of Fractures in Naturally Fractured Reservoirs. Paper SPE 63266 presented at the SPE Annual Technical Conference and Exhibition, Dallas, 1–4 October.

Warren, J.E., and Root, P.J. 1963. The Behavior of Naturally Fractured Reservoirs. *SPEJ* **3** (3): 245–255. SPE-426-PA.

White, N.J., Jackson, J.A., and McKenzie, D.P. 1986. The relationship between the geometry of normal faults and that of the sedimentary layers in their hanging walls. *J. Struct. Geol.* **8:** 879–909.

Wilkinson, W.M. 1953. Fracturing in Spraberry reservoir, West Texas. *AAPG Bull.* **37:** 250–265.

World Stress Map Project. 2004. *Stress Map of Eastern USA*, Heidelberg Academy of Sciences and Humanities, U. of Karlsruhe.

Yose, L.A., Brown, S., Davis, T.L., Eiben, T., Kompanik, G.S., and Maxwell, S.R. 2001. 3-D geologic model of a fractured carbonate reservoir, Norman Wells Field, NWT, Canada. *Bull. Cdn. Pet. Geol.* **49:** 86–116.

Zoback, M.D., Barton, C.A., Brudy, M. et al. 2003. Determination of stress orientation and magnitude in deep wells. *Intl. J. Rock Mechanics & Mining Sci.* **40:** 1049–1076.

Index

A

Absolute open flow, 3-4
Acoustic image logs, 38-39
Altamont-Bluebell oil field, 20
American Association of Petroleum Geologists
  (AAPG), 20, 38, 51
Amott index, 61-63
Analogs, 52-53
Anisotropy, 8, 13, 61, 96
  Ekofisk field and, 85-89
  fracture intersections and, 67
  interference testing and, 79
  permeability and, 82
  Spraberry trend area and, 11-12
  tracers and, 79-80
Aperture, 20-22, 37, 41
  density and, 47-50
  false positive readings and, 42
  photoelectric effect (PEF) logs and, 43
  porosity logs and, 43
  spatial organization and, 46-47
Appalachian Plateau, 19
Arches National Park, Utah, 19
Austin Chalk, 53
Aztec formation, 25

B

Bach Ho field, Vietnam, 26
Basement rock, 26-27
Bit chatter, 9
Bit drop, 9
Black Hills, South Dakota, 27
Borehole. See Wells
Breccia, 30
Brittleness, 32-33
Bubble maps, 4-5
Bubblepoint, 10-11
Buildup test, 56

C

Calcite, 26
Caliper logs, 45-46
Cantarell field, 63
Capillary/gravity ratio, 63
Capillary imbibition
  Ekofisk field and, 91-92
  Midale field and, 74
  simulation models and, 68-69
  Spraberry trend area and, 76, 82-84
  wettability and, 61-67
Capillary pressure, 63
Carbonates, 62
Casing-shoe leakoff test (minifrac), 33, 78-79
Cave maps, 26
"Characterization of Faulting and Fracturing in
  Ekofisk Field From Seismic, Core and Log Data"
  (Dangerfield, Knight, and Farrell), 88
Clay, 28
CO_2 injection
  Midale area and, 71-75
  Spraberry trend area and, 75, 77, 84

Compressibility, 56-57
Connectivity, 8, 65-67
Constant-flow test, 55
Contoured stereonet, 47
Core logs, 36-38
  density and, 47-50
  Ekofisk field and, 91
  false positive readings and, 42
  porosity and, 47-50
Cracks, 18
Cretaceous age, 85
Curvature, 31-32

D

Danian age, 85
Dual-porosity model, 68
Death Valley National Park, 31
Deformation bands, 18, 29
Delaware basin, Texas, 16
Density
  DFN models and, 60, 69
  discriminant analysis and, 51
  fieldwide model and, 52
  fracture interpolation and, 47-52
  neural-net-based routines and, 51
  porosity logs and, 43
  shape factor and, 52, 66
DFN models, 60, 69
Diagenetic alteration, 26
Dimensionless time, 62
Dipmeter logs, 46-47
Discriminant analysis, 51
Displacement history, 25
Dolomites, 26, 73-74
Downhole camera-compass assemblies, 37
Drawdown test, 54, 56-57
Drivers, 52
Dual-permeability/dual-porosity (DKDP) model, 68
Dual-porosity model, 74
  fluid flow and, 62, 65-68
  NFR characterization and, 54, 56-57
Dynamic data
  multiple-well tests and, 57-58
  single-well tests and, 54-57

E

Economic issues, 4
Efficiency
  heterogeneity and, 2-9, 59-60
  Midale field and, 77
  Productivity Improvement Factor (PIF) and, 3
  spatial organization and, 46-47
  sweep, 77
Ekofisk field
  case history of, 84-97
  core log of, 91
  enhanced oil recovery and, 90
  fluid properties of, 85
  fracture analysis of, 85-89
  free water level variation in, 92
  gas/oil ratio (GOR) and, 85

geology of, 84-85
imbibition data for, 91-92
interference tests and, 95-96
permeability and, 85-89, 96-97
porosity and, 89, 93-94
pressure system identification and, 92
radioactive tracer observation and, 97
reservoir properties of, 85
structural position of, 89
Tor formation and, 85, 90, 92
water injection and, 90-95
well tests and, 90, 92, 94, 97
wettability and, 94
Elk basin, Wyoming, 24
En echelon surfaces, 25
Equations
  capillary/gravity ratio, 63
  capillary imbibition, 62
  dimensionless time, 62
  flow capacity index (FCI), 6
  fracture density, 48
  fracture porosity, 49, 57
  interporosity flow coefficient, 57
  sampled interval volume, 48
  shape factor, 66
  storativity ratio, 56
  three-dimensional curvature, 31
  transfer function, 65
Erosional unloading, 20, 33
Exfoliation joints, 27

F

Factor productivity index (FPI), 6
False positive readings, 42
Faults, 16
  breccia and, 30
  definition of, 18
  deformation bands and, 29
  Ekofisk field and, 85-89
  extension fractures and, 15
  fluid flow and, 28
  gouge material and, 28
  induced fractures and, 30-31
  juxtaposition maps of, 28
  large-scale, 85
  orthogonal orientation of, 15
  seal analysis and, 27
  shapes of, 28
  as shear fractures, 15, 27-28
  smear and, 28
Flow
  absolute open, 3-4
  breakthrough and, 61
  capillary imbibition and, 61-67
  constant-flow test and, 55
  directional, 12-13
  dual porosity and, 54, 56-57
  faults and, 27-31
  gas-gravity drainage and, 63-67
  interporosity flow coefficient and, 57
    (see also Porosity)
  joints and, 22
  linear behavior and, 54-55
  matrix/fracture interaction and, 61-64
  multiple-well tests and, 57-58
  network connectivity and, 65-67
  pore-fluid overpressure and, 32

preferential orientation and, 11
production logging tool (PLT) and, 7, 35, 44
radial, 55
shape factor and, 66
simulation models and, 60-64, 67-69
single-well tests and, 54-57
stylolites and, 29-30
transfer function and, 61-62, 65-66
wettability and, 61-67
Yates field study and, 64-67
Flow capacity index (FCI), 3-4, 6-7, 90
Flowmeters, 7
Fluid breakthrough, 61
Folds, 31-32
"Fractured Jargon" (Narr, Engelder, Lacazette, and
  Willemse), 18
Fracture productivity index (FPI), 6
Fractures, 20-22
  aperture measurement and, 37, 41
  capillary imbibition and, 61-67
  core logs and, 36-38
  density and, 47-52
  detecting in well, 35-42
  development timing of, 37
  false positive readings and, 42
  gas-gravity drainage and, 63-67
  gas shows and, 45
  geochemical modification and, 37
  height measurement and, 37, 41-42
  image logs and, 36, 38-42
  indirect indicators and, 42-46
  lost circulation and, 44-45
  matrix, 4-12, 56-57, 61-67
  mechanical indications of, 45-46
  mineralization and, 38
  mudlogs and, 45
  multiple-well tests and, 57-58
  network connectivity and, 65-67
  orientation and, 37 (see also Orientation)
  PEF logs and, 43
  porosity logs and, 43, 47-52
  production logging tool (PLT) and, 44
  representative elemental volume (REV) and, 50
  rose diagrams and, 46-47, 77
  set assignment and, 46
  shape factor and, 66
  single-well tests and, 54-57
  slickenslides and, 38
  stereonets and, 46-47
  Stoneley wave logs and, 42
  stylolites and, 29-30, 85-89
  tectonic, 85-89
  validation of, 42-46
  weathering and, 38
  wettability and, 61-67
  wireline-based sensing of, 36
  Yates field study and, 64-67
Free water level (FWL), 92
Frequency distribution plot, 4
Fringe, 25

G

Gas
  capillary imbibition and, 61-63
  Productivity Improvement Factor (PIF) and, 3
  wettability and, 61-62
Gas-gravity drainage, 63-67

Gas injection. *See* CO_2 injection
Gas/oil ratio (GOR), 8, 11, 85
Gas shows, 45
Gaussian curvature, 31-32
Geology
  breccia and, 30
  brittleness and, 32-33
  characterization and, 35-60
  contemporary stress fields and, 33-34
  core logs and, 36-38
  Cretaceous age and, 85
  Danian age and, 85
  data synthesis and, 58-60
  density and, 47-52
  Earth's crust and, 15
  Ekofisk field and, 84-85
  faults and, 15-16, 27-31
  folds and, 31-32
  fracture concepts and, 15-17
    (*see also* Fractures)
  image logs and, 36, 38-42
  joints and, 15-27
  lithologic controls and, 32-33
  Maastrichtian age and, 85
  multiple-well tests and, 57-58
  orientation and, 37 (*see also* Orientation)
  Paleocene age and, 85
  pore-fluid overpressure and, 32
  single-well tests and, 54-57
  stress history and, 33
  stylolites and, 29-30, 85-89
  well fracture detection and, 35-42
  Yates field and, 63-67
Gouge, 28
Gravity
  capillary forces and, 63-64
  drainage and, 63-67
  Midale field and, 73
  simulation models and, 68
Gulf of Mexico, Texas, 29

H

Hackle marks, 25
Haft-Kel field, 63
Hatch patterns, 32
Heterogeneity
  bubble maps and, 4-5
  flow capacity index (FCI) and, 3-4, 6-7
  fracture importance and, 4
  productivity and, 2-9, 59-60
Histograms, 6
Humble Oil Co., 76

I

Image logs, 36
  acoustic, 38-39
  density and, 47-50
  false positive readings and, 42
  fracture aperture and, 41
  fracture height and, 41-42
  mud and, 40
  orientation and, 38-42
  porosity and, 47-50
  resistivity-based, 38-39
  software for, 40
Initial potential (IP) distribution, 3-4, 11-12
Interference tests, 79, 81, 95-96

Interporosity flow coefficient, 57
Iodine-125, 97

J

Jalama Beach, California, 33
Joints, 16
  aperture and, 20-22
  basement rock and, 26-27
  definition of, 18
  diagenetic alteration and, 26
  distribution curves and, 22
  en echelon surfaces and, 25
  erosional unloading and, 20
  exfoliation, 27
  as extension fractures, 15, 17
  fracture geometry and, 20-22
  fringe and, 25
  height/length and, 20-22
  layering angle and, 17
  orthogonal orientation and, 15, 17, 20-22
  relative age and, 17-18
  sedimentary rock and, 21-22, 26-27
  set of, 17
  sheeting, 28
  spacing and, 22
  surface markings and, 22-26
  system of, 17
  veins and, 26
Journal of Structural Geology, 29
Judith Creek Sandstone, 24

K

Kalaydijian, 65
Kazakhstan, 43
Kinematic aperture, 37
Kinematic history, 26

L

Lake Maracaibo, Venezuela, 30
La Paz field, Venezuela, 26
Listric shapes, 28
Lithology, 30
  analogs and, 52-53
  density and, 47-52
  fracture occurence and, 32-33
Lost circulation, 44-45

M

Mara field, Venezuela, 26
Marly zone, 73-74
Masstrichtian age, 85
Matrix/fracture interaction, 56-57
  capillary imbibition and, 61-67
  flow capacity index (FCI) and, 4, 6-7
  gas-gravity drainage and, 63-67
  Spraberry trend area and, 9-12
  transfer function and, 61-62, 65-66
  wettability and, 61-67
Matrix porosity, 51
Midale field, Saskatchewan, 71-77
Mineralization, 37
  growth textures and, 26
  natural fractures and, 38
  secondary, 1-2
  spatial organization and, 46-47
Minifrac test, 33, 78-79
Miocene sandstone, 29

Monterrey, Mexico, 24
Mud
  gas shows and, 45
  lost circulation and, 44-45
  photoelectric effect (PEF) logs and, 43
  resistivity-based image logs and, 40
Multiple-regression methods, 51
Multiple-well tests, 57-58
Multiwell interference testing, 79

N

National Park Service, 27
Naturally fractured reservoirs (NFRs).
    See also Specific field
  abundance of, 1
  basement rock and, 26-27
  breccia and, 29-30
  characterization of, 1-2, 35-60, 62, 65-66
  cracks and, 18
  crust outcrops and, 15
  curvature and, 31-32
  data synthesis for, 58-60
  defined, 1-2
  deformation bands and, 18
  diagenetic alteration and, 26
  dynamic data and, 53-58
  evaluation prudence for, 1
  faults and, 15-16, 27-31
  field analog production and, 52-53
  fluid flow in, 61-69
  folds and, 31-32
  fracture concepts and, 15-17
    (*see also* Fractures)
  fringe and, 25
  geology of, 15-34
  indirect indicators and, 42-46
  joints and, 15-27
  lithologic controls on occurence and, 32-33
  multiple-well tests and, 57-58
  orientation and, 46-50 (*see also* Orientation)
  outcrop analogs and, 53
  pore-fluid overpressure and, 32
  predictability issues and, 30
  productivity heterogeneity and, 2-9
  Productivity Improvement Factor (PIF) and, 3
  recognizing, 1-14
  simulation of, 54, 56-57, 61-69
  single-well tests and, 54-57
  spatial organization of, 46-50
  Spraberry trend area and, 9-13
  stress fields and, 33-34
  stylolites and, 29-30, 85-89
  surface markings and, 22-26
  understanding of, 13-14
  veins and, 18, 26
  well fracture detection and, 35-42
Network connectivity, 8, 65-67
Neural-net-based routines, 51
Norman Wells field, Northwest Territories, 59

O

Oil
  enhanced recovery and, 90
  gas/oil ratio (GOR), 8, 11, 85
  original oil in place (OOIP), 74-75, 84
  Productivity Improvement Factor (PIF) and, 3
  water/oil ratio (WOR), 8

Oil-wet carbonates, 62
Orientation
  core logs and, 37
  density and, 47-50
  dipmeter logs and, 46-47
  downhole camera-compass assemblies and, 37
  geology models and, 37
  image logs and, 38-42
  joints and, 15, 17, 20-22
  paleomagnetic field and, 37
  porosity and, 47-50
  preferential, 11
  rose diagrams and, 46-47
  spatial organization and, 46-50
  Spraberry trend area and, 76
  stereonets and, 46-47
Original oil in place (OOIP), 74-75, 84
Outcrops
  analogs and, 53
  faults and, 15-16, 27-31
  joints and, 15-17
Overperformance, 6

P

Paleocene age, 85
Pay zones, 11
Permeability, 59
  capillary imbibition and, 61-63
  DFN models and, 60
  directional, 11
  Ekofisk field and, 84-89, 96-97
  enhancement of, 61
  matrix-fracture interaction and, 61-64
  relative, 65
  shape factor and, 66
  simulation models and, 60-64, 67-69
  Spraberry trend area and, 10-11, 76
  vertical, 97
  well-test vs. core, 11
  wettability and, 61-63
  Yates field and, 63-67
Permeability anisotropy, 61
Permeability thickness (kh), 3, 6-7, 54-57
Permian Basin, 64-67
Photoelectric effect (PEF) log, 43
Pipes, 65
Plumose structures, 25
Pollard, D.D., 22
Pore-fluid overpressure, 32
Porosity, 8, 43
  density and, 47-52
  DFN models and, 60
  dual, 54, 56-57, 62, 65-68, 74
  Ekofisk field and, 89, 93-94
  fracture, 49, 57
  interporosity flow coefficient and, 57
  matrix, 51
  Midale field and, 71-75
  NFR characterization and, 47-50
  simulation models and, 60-64, 67-69
  transfer function and, 65-66
  vuggy, 42, 68
Production logging tool (PLT), 7, 35, 44
Productive height (h), 7
Productivity
  80/20 rule and, 4
  bubble maps and, 4-5

cumulative, 3-4
Ekofisk field and, 84-97
expected vs. observed, 4, 6
factor productivity index (FPI) and, 6
flow capacity index (FCI), 3-4, 6-7, 90
fracture-enhanced, 2-3
heterogeneity and, 2-9
Midale field and, 71-75
monthly production and, 4
overperformance and, 6
spatial distribution of, 3
Spraberry trend area and, 9-12, 75-84
Productivity Improvement Factor (PIF), 3
Productivity index (PI), 3-4, 66
P-waves, 9

Q–R

Quartz, 26
Radioactive tracer tests, 72, 97
Rang Dong field, Vietnam, 26
Regional joint set, 53
"Relationship Between the Geometry of Normal
  Faults and That of the Sedimentary Layers in
  their Hanging Walls, The" (White, Jackson,
  and McKenzie), 29
Representative elemental volume (REV), 50
Resistivity, 8, 38-39
Rocky Mountain National Park, Colorado, 28
Rose diagrams, 46-47, 77

S

Salt tests, 72
San Andreas formation, 64-65
San Onofre, California, 29
Sedimentary rock, 21-22, 26-27
Shape factor, 52, 66
Shear fractures. See Faults
Shear waves, 9
Sheeting joints, 28
Sierra Madre Oriental, Mexico, 37
Siltstone, 25
Simulation models
  3D modeling software and, 46-47, 53
  capillary imbibition and, 61-67
  DFN, 60, 69
  dual-permeability/dual-porosity, 68
  dual porosity, 54, 56-57, 62, 65-68, 74
  matrix-fracture interactions and, 61-64
  single effective medium, 68-69
Single-well tests, 54-57
Skov, A.M., 76
Skull Creek anticline, 23
Slickensides, 38
Smear, 28
Sonic porosity logs, 43
Spacing distributions, 22
Spikes
  caliper logs and, 45-46
  mudlog and, 45
  photoelectric effect (PEF) logs and, 43
Spraberry trend area
  capillary imbibition and, 76
  case history of, 75-84
  CO_2 injection and, 75, 77, 84
  complicated fractures of, 76
  description of, 9-10
  directional fluid movement and, 11-12

directional permeability and, 11
Driver unit, 76
fracture intersections and, 67
gas/oil ratio (GOR) increase and, 11
horizontal core and, 78
injection profiles and, 81-82
Midkiff area of, 76
minifrac test and, 78-79
multiwell interference testing and, 79, 81
O'Daniel plot, 75, 77-78
orientation and, 76
permeability and, 11, 76
rapid production decline in, 9-11
step-rate injection test (SRIT) and, 78-79
tracer survey and, 79-81
water injection and, 75-76, 82-84
Step-rate injection test (SRIT), 78-79
Stereonets, 46-47
Stoneley-wave logs, 41-42
Storativity ratio, 56
Stress
  casing-shoe leakoff test and, 33
  contemporary field for, 33-34
  cooling induced, 27
  history and, 33-34
  image logs and, 42
  thermal, 27, 33
*Structural and Tectonic Modeling and Its
  Application to Petroleum Geology,* 88
Stylolites, 29-30, 85-89
Surface markings, 22-26
Sweep efficiency, 77

T

Tadpole logs, 46-50
Temperature logs, 44
Tengizchevroil, 43
Tor formation, 85, 90, 92
Tracer surveys, 58, 72
  Ekofisk field and, 97
  Spraberry trend area and, 79-81
Training, 51
Transfer function, 61-62, 65-66
Tritium, 97

U

Uinta basin, Utah, 51
University of Texas Bureau of Economic
  Geology, 37

V

Valley of Fire State Park, Nevada, 25
Veins, 18, 26
Vertical seismic profile (VSP), 9
Viscosity, 62
Viscous stripping, 73
Vugs, 42, 68
  Midale field and, 71-75

W

Washout zones, 42
Water injection
  Ekofisk field and, 90-95
  Midale field and, 71-77
  Spraberry trend area and, 75-76, 82-84
Water/oil ratio (WOR), 8
Weathering, 38

Weber formation, 20
Well *kh* ratio, 54-57
Wells
  core logs and, 36-38
  data synthesis and, 58-60
  density and, 47-50
  dual porosity and, 54, 56-57
  dynamic data and, 53-58
  Ekofisk field and, 84-97
  fracture detection in, 35-42
  image logs and, 36, 38-42
  indirect indicators and, 42-46
  lost circulation and, 44-45
  Midale field and, 71-75
  multiple-well tests and, 57-58
  porosity and, 47-50
  production logging tool (PLT) and, 7, 35, 44
  single-well tests and, 54-57
  Spraberry trend area and, 75-84
  vertical, 38
Wettability, 61-62
  Ekofisk field and, 94
  gas-gravity drainage and, 63-67
  Midale field and, 74
  transfer function and, 65-66
Whiskey Gap, Wyoming, 20
Wind Cave, 27
Windgate sandstone, 19
World Stress Map Project, 34

X-Z

Yates field, Texas, 63-67